Springer Theses

Recognizing Outstanding Ph.D. Research

Aims and Scope

The series "Springer Theses" brings together a selection of the very best Ph.D. theses from around the world and across the physical sciences. Nominated and endorsed by two recognized specialists, each published volume has been selected for its scientific excellence and the high impact of its contents for the pertinent field of research. For greater accessibility to non-specialists, the published versions include an extended introduction, as well as a foreword by the student's supervisor explaining the special relevance of the work for the field. As a whole, the series will provide a valuable resource both for newcomers to the research fields described, and for other scientists seeking detailed background information on special questions. Finally, it provides an accredited documentation of the valuable contributions made by today's younger generation of scientists.

Theses are accepted into the series by invited nomination only and must fulfill all of the following criteria

- They must be written in good English.
- The topic should fall within the confines of Chemistry, Physics, Earth Sciences, Engineering and related interdisciplinary fields such as Materials, Nanoscience, Chemical Engineering, Complex Systems and Biophysics.
- The work reported in the thesis must represent a significant scientific advance.
- If the thesis includes previously published material, permission to reproduce this must be gained from the respective copyright holder.
- They must have been examined and passed during the 12 months prior to nomination.
- Each thesis should include a foreword by the supervisor outlining the significance of its content.
- The theses should have a clearly defined structure including an introduction accessible to scientists not expert in that particular field.

More information about this series at http://www.springer.com/series/8790

Supervisor's Foreword

In February 2013, I relocated from Singapore to Australia to serve as the Director of Queensland Micro- and Nanotechnology Centre (QMNC) at Griffith University in Queensland, Australia. QMNC has a unique capability in silicon carbide technology, which up to this time has been focused mainly on microelectronics. In 2013, the field of Micro Electromechanical Systems (MEMS) has been established for more than three decades and MEMS products have found their way into a wide range of consumer products such as the mobile phone. However, the majority of MEMS devices are made of silicon. In August 2013, Dr. Dzung Dao, who joined the centre two years earlier, introduced me to Hoang-Phuong Phan, a new PhD student, who just arrived from the University of Tokyo. Phan presented the past MEMS work from his master thesis, and I was impressed with the depth, the rigor of the work, as well as the excellent research skills he already acquired. Indeed, Phan has an illustrious academic track record prior joining Griffith University. In 2004, Phan won the third price in the National Mathematical Olympiad of Vietnam, which allowed him to gain direct admission to Hanoi University of Science and Technology. Being one of the best in his cohort, he then obtained a scholarship from the Japanese government to study at the prestigious University of Tokyo. After obtaining his Bachelor of Engineering degree in 2011, he received another scholarship from the Japanese government for postgraduate study toward the Master of Engineering degree. Phan continued his research at Griffith University as a member of QMNC. He is fluent in Vietnamese, Japanese and English.

The present monograph is the result of Phan's research over a span of 30 months. With his solid background in MEMS, Phan was able to complete a high-quality Ph.D. thesis on the piezoresistive effect of silicon carbide, which opens up a range of MEMS applications for harsh environment. Within this relatively short period, Phan authored and co-authored an impressive list of 18 journal papers in high-impact journals with over 250 citations to date. This book first provides an overview on the fundamentals and the state-of-the-art of piezoeresistive effects in silicon carbide. The overview is followed by the theoretical consideration of piezoresistive effect of p-type 3C-SiC. The remaining chapters of the book report the fabrication and the characterisation of p-type 3C-SiC piezoresistors, especially nanocrystalline films and nanowires. Phan's work not only gave him the Ph.D. degree, but also opened up at QMNC a new research direction on SiC based sensors and actuators, which is currently followed up by other Ph.D. students at the centre.

Readers could find in this book an exemplary research approach for the investigation of piezoresistive effect in wide-band-gap materials, particularly of the different poly types of silicon carbide. The introduction to silicon carbide is enjoyable to read and it eases the understanding of the subsequent chapters. Although this book originates from a Ph.D. thesis, the content suits well as an introductory material for engineers and postgraduate researchers. I hope that this book is the first of more to come from Phan's future career. The book was proof read and checked carefully. However, errors and mistakes are unavoidable. We welcome any feedback and correction from the readers.

Brisbane, Australia
December 2016

Prof. Nam-Trung Nguyen

Abstract

Silicon carbide (SiC) is an excellent material for electronic devices operating at high temperatures, thanks to its large energy band gap, superior mechanical properties and extreme chemical inertness. Among numerous poly types of SiC, the cubic single crystal, which is also well-known as 3C-SiC is the most promising platform for MEMS applications, as it can be epitaxially grown on a Si substrate with diameters up to several hundred millimeters. This capability makes 3C-SiC compatible with the conventional Si based micro/nano processing and also cuts down the cost of SiC wafers. The investigation into the piezoresistive effect in 3C-SiC is of significant interest for the development of mechanical transducers such as pressure sensors and strain sensors used for controlling combustion and deep well drilling. Although a number of studies have been focused on the piezoresistive effect in n-type 3C-SiC, 4H-SiC and 6H-SiC; the piezoresistance in p-type 3C-SiC has been rarely reported. This book presents the piezoresistance in p-type 3C-SiC which was investigated using experimental characterization and theoretically analysis. The gauge factor, the piezoresistive coefficients in two-terminal and four-terminal resistors, the comparison between single crystalline and nanocrystalline SiC, along with the temperature dependence of the piezoresistive effect in p-type 3C-SiC were also addressed. In addition, the piezoresistive effect of top-down fabricated SiC nanowires were also investigated, showing a high sensitivity in nanowires employing an innovative nano strain-amplifier. The large gauge factors of the p-type 3C-SiC at both room temperature and high temperatures found in this study indicated that this poly type is feasible for the development of mechanical sensing devices operating in harsh environments with high temperatures.

Acknowledgements

I would like to express my gratitude to many people who gave me great support, encouragement and guidance during my Ph.D. course.

Firstly, I would like to express my appreciation to my principal supervisors: Dr. Dzung Dao and Prof. Nam-Trung Nguyen. It is my privilege to work under the supervision of Dr. Dzung Dao who is always ready and willing to give me chances to discuss my research. Dr. Dzung Dao is an expert in the field of the piezoresistive effect in semiconductors. He was one of the main reasons that made me to decide doing my Ph.D. at Griffith University. From Dr. Dzung Dao I have learned a huge amount of knowledge related to the piezoresistive effect, as well as the physics of semiconductors. I am also very lucky to have worked under the supervision of Prof. Nam-Trung Nguyen. It was very helpful and enjoyable to discuss with him about my research ideas and problems that I need to solve. From Dr. Dzung Dao and Prof. Nam-Trung, I could not only gain more insight, but also learn how to be a good researcher, as well as a good person. Once a gain I deeply acknowledge everything they have done for me.

I am deeply grateful to Prof. Sima Dimitrijev who has helped me to profoundly understand the electrical properties of semiconductor, especially SiC. I was amazed with his profound knowledge. I hope that I could learn as much as possible from him in the future. I also would like to thank Dr. Philip Tanner, Dr. Li Wang, Dr. Jisheng Han, Lin Zhang, Glenn Walker, and Alan Iacopi for helping me in designing and fabricating the samples and to conduct experiments in Queensland Micro- and Nanotechnology Centre. Without their help, it would be very hard for me to carry out my study on the piezoresistive effect in SiC. I am further indebted to Dr. Yong Zhu, who is also working at the MEMS group at Gold Coast campus. The times I discussed with him about MEMS devices were very enjoyable.

I am also thankful to Prof. Namazu's group at University of Hyogo, Japan for their help in the fabrications of the released/transferred SiC samples using focused ion beam. I would like to acknowledge the support from my lab-mates and friends who are also doing their Ph.D. at Griffith University. I am thankful for their encouragement and their helps in sample preparation and experiments.

I also appreciate the support from my family, who always give me the trust and help whenever I need. Thanks dad for always being brave, positive and optimistic even though suffering several serious diseases for a long time. Your love gave me great energy, motivation and encouragement. You will always be in my heart. This one is dedicated to you, dad.

Last but not least, I want to thank Griffith University for granting the GUPRS and GUIPRS scholarships, as well as giving me a great opportunity to carry out my Ph.D. research here. Griffith University has a wonderful environment and advanced facilities for every student who wants to conduct high-impact research. I am very happy and lucky to be a part of this university.

Contents

Abbreviations

AFM	Atomic Force Microscopy
APCVD	Atmospheric Pressure Chemical Vapor Deposition
CVD	Chemical Vapor Deposition
GF	Gauge Factor
HMCVD	Hot Mesh Chemical Vapor Deposition
LPCVD	Low Pressure Chemical Vapor Deposition
MEMS	Micro Electro Mechanical Systems
MWCVD	Micro Wave Chemical Vapor Deposition
PECVD	Plasma Enhanced Chemical Vapor Deposition
SAED	Selected Area Electron Diffraction
SEM	Scanning Electron Microscopy
TEM	Transmission Electron Microscopy
XRD	X-Ray Diffraction

Physical Constants

Boltzmann constant	$k_B = 1.38 \times 10^{-23}$ m^{2}kgs^{-2}K^{-1}
Stefan Boltzmann constant	$\lambda = 5.670 \times 10^{-8}$ Wm^{-2}K^{-4}
Planck's constant	$h = 6.626 \times 10^{-34}$ m^{2}kgs^{-1}
Unit charge	$q = 1.602 \times 10^{-19}$ C
Young's modulus of 3C-SiC	E_{SiC} = 330 GPa
Young's modulus of Si ([100] direction)	$E_{Si,[100]}$ = 130 GPa
Young's modulus of Si ([110] direction)	$E_{Si,[110]}$ = 169 GPa

Symbols

GF	Gauge factor (–)
GF_l	Longitudinal gauge factor (–)
GF_t	Transverse gauge factor (–)
$\pi_{11}, \pi_{12}, \pi_{44}$	Fundamental piezoresistive coefficients (Pa^{-1})
π_l	Longitudinal piezoresistive coefficient (Pa^{-1})
π_t	Transverse piezoresistive coefficient (Pa^{-1})
π_s	Shear piezoresistive coefficient (Pa^{-1})
σ	Stress (MPa)
σ_c	Electrical conductivity (S/m)
σ_{Si}	Electrical conductivity of Si (S/m)
σ_{SiC}	Electrical conductivity of SiC (S/m)
m_h^*	Hole effective mass (kg)
τ_h	Hole mean free path (m)
σ_{ij}	Stress tensors (MPa)
ε_{SiC}	Strain of SiC resistor (–)
ρ	Resistivity of SiC (Ωcm)
ρ_{ij}	Resistivity tensors of SiC (Ωcm)
R	Resistance of SiC (Ω)
t_{SiC}	Thickness of the SiC layer (m)
t	Thickness of the Si layer (m)
l	Length of Si beam (m)
w	Width of Si beam (m)
$\frac{\Delta\rho}{\rho}_{L,[100]}$	Relative resistivity change in longitudinal [100] direction (–)
$\frac{\Delta\rho}{\rho}_{T,[100]}$	Relative resistivity change in transverse [100] direction (–)
$\frac{\Delta\rho}{\rho}_{L,[110]}$	Relative resistivity change in longitudinal [110] direction (–)
$\varepsilon_{L,[100]}$	Strain of SiC resistor aligned in longitudinal [100] direction (–)
$\varepsilon_{T,[100]}$	Strain of SiC resistor aligned in transverse [100] direction (–)
$\varepsilon_{L,[110]}$	Strain of SiC resistor aligned in longitudinal [110] direction (–)

Chapter 1
Introduction and Literature Review

1.1 Research Background of the Piezoresistive Effect in SiC

The piezoresistance is defined as the change in electrical resistance of a material under external mechanical strain or stress, which was discovered by Smith in 1954 [1]. Since then, a great number of research works have been relentlessly carried out to elucidate the phenomenon in numerous materials. Besides fundamental investigation, applications of the piezoresistive effect in semiconductors can be found in numerous Micro Electro Mechanical Systems (MEMS) sensors, thanks to its superior properties, including device miniaturization, simple readout circuit, and low power consumption [2, 3], compared to other sensing technologies (e.g. electrostatic, piezoelectric and optical). Among various semiconductors, silicon (Si) material is the most favorable platform for the development of piezoresistive transducers owing to the significant magnitude of effect, advancement in micro and nano machining resulting in mass production, as well as low cost [4–6]. Pressure, tactile, inertia and AFM (atomic force microscopy) cantilevers are the most typical applications utilizing the piezoresistive effect of Si [7–11].

In fuel combustion industries including automobile and aerospace, there is an increasing demand for monitoring in real time the pressure level of the engine chambers at elevated temperatures. The measurement of the temperature and pressure inside combustion chambers can be employed to evaluate and optimize the performance of engines, and thus could improve the efficiency of the combustion process [12]. Additionally, real-time monitoring the mechanical impact on hot sections of turbines also play an vital role to predict and avoid any possible failure occurred in engines [13]. Therefore, the development of mechanical sensing devices which can withstand hostile conditions such as high temperatures and increasing corrosion is an increasingly important concern [14–16].

Silicon is not the material of choice for such applications due to its relatively low energy band gap (1.12 eV), which prevents the Si based piezoresistive MEMS sensors to below 200 °C [17, 18]. As a result, the piezoresistance of other wide band gap materials such as silicon carbide (SiC), gallium nitride (GaN), and diamond

H.-P. Phan, *Piezoresistive Effect of p-Type Single Crystalline 3C-SiC*,
Springer Theses, DOI 10.1007/978-3-319-55544-7_1

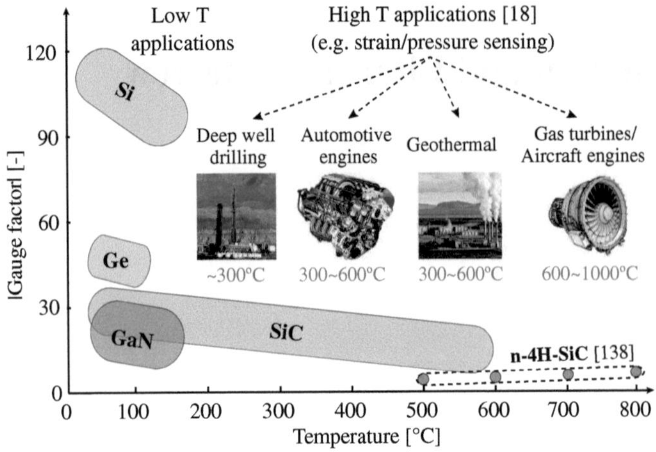

Fig. 1.1 The piezoresistance of some common materials such as silicon [1], germanium [26], gallium nitride [27], and silicon carbide [14]. Silicon based piezoresistive devices are favorable for low-temperature applications, including strain gauge, pressure sensors, inertial measurement unit, and cantilevers which typically operates below 200 °C. On the other hand, gallium nitride (GaN) [27, 28], and Silicon carbide [14] are suitable materials for MEMS transducers worked at higher temperatures as well as hostile condition such as deep well-drilling, combustion chambers, and space exploration [12]

like carbon (DLC) is of interest for these niche applications at high temperatures (Fig. 1.1) [14, 19]. Compared to other common large band gap materials such as DLC and GaN, SiC possesses several advantages for MEMS transducers operating in harsh environments due to its wide energy band gap, large Young's modulus, along with extreme chemical inertness [20, 21]. Furthermore, the commercial availability of large-scale SiC wafers with diameter above 4 in. [23, 24], the rapid growth in the market size of SiC electronics [22], as well as advancement in MEMS/NEMS fabrication technologies [25] also bring great potential to SiC over other wide band gap semiconductors.

The investigation into the piezoresistance of SiC has been relentlessly carried out since the 90's decades, both by theoretical analysis and experimental work, along with the demonstration of numerous piezoresistive SiC-based sensors [14]. This chapter gives an overview on SiC material characteristics, its MEMS processing compatibility, and particularly, the state of the art work on the piezoresistive effects in SiC for MEMS transducers. Furthermore, the applications utilizing the piezoresistance of SiC are also presented. From the literature review, the research gaps as well as the aims of this dissertation are presented at the end of this chapter.

1.2 Silicon Carbide as a Material for MEMS Applications

1.2.1 Crystalline Structure and Physical Properties

Silicon carbide exists in three poly types, which are single crystal, poly crystal, and amorphous. In crystalline SiC, each Si atom is joined to four other carbon atoms by covalent bonds, forming tetrahedrons. Silicon carbide exhibits a one-dimensional polymorphism, that results in different poly-types. These SiC poly types are distinguished from one another by the stacking sequence of each Si-C bilayer [29]. There are over 200 crystalline structures which are generally categorized into α-SiC and β-SiC. β-SiC, which is commonly known as 3C-SiC, is the only cubic crystalline structure of SiC, whereas the most common α-SiC crystals are the hexagonal poly types including 2H-SiC, 4H-SiC, and 6H-SiC [30].

Table 1.1 lists the properties of the most common SiC single crystals (3C-SiC, 4H-SiC, and 6H-SiC) in comparison to the conventional MEMS/NEMS material (Si), and other wide band gap semiconductors such as GaN diamond and diamond [20, 31, 32]. Silicon carbide has advantages of the wide band gap ranging from 2.3 eV to 3.4 eV, high carrier mobilities and high breakdown voltage which are all desirable characteristics for high-power electronics as well as high-temperature MEMS applications [33, 34].

Silicon carbide is an excellent material for high temperature transducers, not only due to the large energy band gap, but also owing to the high melting point of over 2500 °C a long with the high thermal conductivity of 5 $\mathrm{Wcm^{-1}K^{-1}}$. Employing the advantages of rapid responses in both heating up and cooling down, numerous SiC thermo-based devices such as fluid flow sensors and gas heaters have been demonstrated [35–37].

Silicon carbide also exhibits superior mechanical properties compared to Si. Thanks to the large hardness of 9.15 (in comparison to 10.0 of diamond), and excellent wear resistance, SiC has been widely employed as a coating material in several micro-machined components to prevent corrosive attacks [20]. Furthermore, SiC possesses higher Young's moduli ranging from 300 GPa to 600 MPa, which is approximately 2–5 times larger than that of Si [38, 39], while the density of SiC is comparable to Si. Consequently, the high ratio between the Young's modulus and the mass density significantly enhances the resonance frequency of SiC devices SiC, making it preferable for ultra high frequency MEMS resonators and oscillators [40]. For instance, SiC resonators having ultra high frequencies exceeding several GHz, and Q factor of above 1 million have been developed [41, 42]. Reviews of SiC based resonators are available elsewhere [43].

Additionally, the superior properties of SiC are also of interest for various bio applications [44]. As such, SiC is an ideal platform for optical bio-sensors owing to its transparency to the visible light. The high corrosion tolerance along with the extreme chemical inertness also make SiC suitable for more reliable and stable in vivo implantation [45]. For instance, micro-needles fabricated using 6H-SiC was applied in tissue monitoring during clinical organ transplantation. These 6H-SiC

Table 1.1 Properties of SiC and other materials

Properties	SiC	Si	Diamond	GaN
Energy gap (eV) [20, 31]	2.3–3.4	1.12	5.5	3.4
Breakdown voltage (V/cm) [20, 48]	4×10^6	3 to 6×10^5	10×10^6	3×10^6
Electron mobility (cm^2/Vs) [20, 48]	1000	1500	2200	900
Hole mobility (cm^2/Vs) [20, 48]	40–100	100–500	1600	150
Young's modulus (GPa) [31, 38]	300 to 500	130–180	1000	200–300
Melting point (°C) [20, 31]	2830[a]	1410	1400[b]	2400
Thermal conductivity (W/cmK) [15, 31]	5	1.5	20	1.3
Chemical Inertness [15]	Excellent	Poor	Good but burn	Good
MEMS compatibility [15]	Good	Excellent	Poor	Fair
Availability/Cost [15]	Fair	Excellent	Poor	Fair

[a]sublimation temperature [20]
[b]phase change temperature [20]

needles noticeably outperformed the conventional Si needles, showing 4-fold higher modulus of rupture, as well as a 10-times extended frequency range [44, 46].

Although there are several large energy band gap semiconductors such as zinc selenide, group III-nitride, and diamond like carbon [30, 47, 48], SiC is among the first choices for harsh environment electronics. That is firstly due to the availability of native SiC substrates and advancement in MEMS micro/nano machining inherited from standard Si technologies [49, 50]. Additionally, in comparison to GaN and AlN, SiC exhibits significantly lower crystal defects and dislocations [16, 18]. For the above-mentioned reasons, it is understandable that more studies have been conducted on SiC than other wide energy band gap semiconductors for high temperature devices [16, 51].

1.2.2 Growth Process of SiC

Hexagonal silicon carbide are formed by bulk growth processes, whereas the cubic poly type is created only by heteroepitaxial chemical-vapor deposition (CVD) on other substrates.

Bulk Growth

Some electronic devices require the use of bulk SiC substrate. As a result, α-SiC (4H and 6H-SiC) is grown for this purpose by bulk-growth processes [20]. The most typical bulk growth of SiC was developed based on the sublimation method proposed by Lely in which the process is usually performed in a quasi-closed graphite crucible [52]. The Lely process was late on modified by Tairov and Tsetkov [53]. The main idea of this bulk process is to sublime SiC powder at high temperatures and then re-crystallize Si- and C-containing gases at a cooler single crystalline seed [53, 54]. Even though the formation of the SiC through vapor–solid and liquid–solid chemical reactions has been known for over 50 years, the growth of single crystalline SiC wafers with the quality equivalent to other conventional semiconductors, such as Si and GaAs, has been a great challenge until recently. The main causes for the difficulty in SiC growth are the high temperature required to crack and forming chemical bonding and the narrow enthalpy difference between the formation of poly-types. To date, the sublimation process is still considered as the most adapted and economically relevant process to produce single crystalline α-SiC [54]. The CVD (chemical vapor deposition) process has recently emerged as a potential alternative to grow bulk SiC wafers, owing to advantages of the continuous supply of the source materials, the direct control of the C/Si ratio, and high-purity gases availability. Generally, in a CVD growth process, a combination of silicon containing source gases (SiH_4, $SiCl_2H_2$, $SiCl_4$) and carbon containing source gases (C_2H_4, C_3H_8, CCl_4, C_6H_{14}) is used. The growth ratio is controlled by the temperature, the flow rate of gases, and the pressure of the reaction chamber. The pressure changes the velocity of gases, which alters the thickness of the boundary layer. The pressure dependency categorizes the CVD process into two types: low pressure, atmosphere pressure CVD.

The requirement for high quality single crystalline SiC films with n-doping and p-doping for electronic devices has propelled the homoepitaxial growth techniques. With the improvement of the growth techniques, high quality and large scale α-SiC wafers of 4 H and 6H-SiC are now available on the market, although the cost is still relatively high.

Hetero-epitaxial Growth

The heteroepitaxial CVD is deployed to grow 3C-SiC on Si substrates, given that 3C-SiC and Si have the same cubic crystalline structure. Consequently, the SiC film has the same orientation as the Si wafers and it can also be deposited on large-diameter wafers. The capability of epitaxial deposition of 3C-SiC offers this poly-type a significant advantage in MEMS applications since the standard MEMS process is applicable and the cost of wafers can be noticeably reduced compared to bulk SiC wafers.

The most common technology employed to deposit single 3C-SiC on Si wafer is LPCVD by simultaneously supplying the silicon precursors and carbon precursors diluted by hydrogen [55, 56]. Initially, the top surface of a Si layer is carbonized by using C precursors, forming a thin layer of 3C-SiC. Subsequently, 3C-SiC films are homo-epitaxially grown on the thin 3C-SiC layer made previously. The temperature of this growth process is usually around 1350 °C, which is very close to the melting point of Si. Consequently, the high deposition temperature may cause other serious effects on SiC films as it accumulates the mismatch in the thermal expansion between SiC and Si, as well as redistributes the dopants in Si. Therefore, great efforts have been made to reduce the deposition temperature. Utilizing alternating-supply or the atomic-layer epitaxy method, the growth temperature has been reduced to around

1000 °C [23, 57]. Zhuang et al. employed microwave CVD (MWCVD) to grow single crystalline 3C-SiC, where the temperature of substrates was lowered to 800 °C [58]. By deploying hot mesh CVD, Yasui et al. have successfully grown 3C-SiC at 700–1000 °C [59]. Along with reducing the growth temperature, it is also important to minimize the interfacial dislocation/defects and other planar defects. An effective method to reduce the density of defects is using selective epitaxial growth (SEG) in which the deposition of 3C-SiC is adjusted to occur only at the exposed substrate areas [60].

Although single 3C-SiC can only be grown on Si substrates, poly-crystalline, nano-crystalline and amorphous 3C-SiC can be deposited on both Si and insulators such as Si_3N_4 and SiO_2 [21]. This gives a significant advantage for these poly-types in MEMS devices, since they can deploy the substrates (e.g. SiO_2, Si_3N_4) for electrical insulation.

Doping of SiC

Thermal diffusion, ion implantation, and in situ doping are conventional methods which are applied to introduce dopants into semiconductors. In SiC, thermal diffusion is impractical, since it requires to be performed at high temperatures above 1800 °C [14]. Ion implantation is possible; however, the requirement of thermal annealing treatment for dopant activation and recrystallization are the main drawbacks of this method [61].

Compared to implantation, in situ doping exhibits uniform dopant profile and the capability of forming an ultra thin piezoresistive layers; therefore it is a suitable method for controlled n-type and p-type doping for sensor fabrication [14, 62]. Ammonia (NH_3) is usually used to create n-type SiC while trymethylaluminium ($Al(CH_3)_3$) or diborane (B_2H_6) are employed to make p-type SiC in the in situ doping process. Table 1.2 shows the ionization energy of common impurities (e.g. N, Al, and B).

1.2.3 MEMS Fabrication Process

To develop MEMS devices, patterning, etching and releasing SiC are required. Therefore, the compatibility between MEMS processes and SiC plays a vital role for fabricating SiC MEMS transducers. SiC movable micro structures can be realized by the following fabrication processes.

Table 1.2 Ionization energy (meV) of impurities in SiC [14]

	3C-SiC	4H-SiC	6H-SiC
Nitrogen	53.6	55	95
Aluminum	260	195	220
Boron	735	655	320

Etching of SiC

Silicon carbide can be patterned by both wet etching and dry etching [21]. Since SiC has a high inter-atomic bonding, it has an excellent inertness, leading to difficulty in etching processes. Molten KOH at 500 °C [63] and orthophosphoric acid heated at 215 °C [64] have been used for the chemical etching of SiC. Nevertheless, since the etching rate is extremely slow and requires high temperatures, this technique is impractical for micro fabrication [64].

Alternative methods use photoelectrical techniques or plasma etching. Photoelectrical etching is known as an effective method for patterning the epilayer of bulk α-SiC wafers, which can selectively etch the top SiC layer without attacking the SiC layer underneath, with a different dopant. Shor et al. developed a wet etching method using ultraviolet illumination in HF solution to selectively remove the top n-type epitaxial 3C-SiC on a p-type 3C-SiC substrate [65].

Dry etching is usually carried out employing a plasma of fluoride gases (e.g. $SF_6, CF_4, CHF_3, NF_3, CBrF_3$) [66, 67]. SiC dry-etching rates have been reported to be several tens of nanometers per minute, which is remarkably lower than Si. However, by combining O_2 and SF_6 gases with a proper mixing ratio, an etching rate as high as 1 μm/min has been demonstrated, indicating the potential of dry etching for patterning thick SiC films [67].

Wafer Bonding Techniques

Wafer bonding can form an electrical insulator under the β-SiC layer or to form α-SiC heteroepitaxial structures. Vinod et al. deployed polysilicon-to-polysilicon bonding to produce single 3C-SiC film on SiO_2 substrate [68]. The fabrication concept of this process is described in Fig. 1.2a. After a single epitaxial 3C-SiC film was deposited on a Si substrate using APCVD, the SiC layer was oxidized to create a 0.2-μm thick SiO_2 layer. In the next step, a polysilicon layer was deposited onto the SiO_2 layer. Another wafer with a polysilicon layer on the top, a SiO_2 layer sandwiched at the middle, and a thick Si handling layer underneath was also prepared. The polysilicon layers were then bonded (Fig. 1.2a, step 3) followed by the removal of the Si substrate, which produced single 3C-SiC on SiO_2 structure (Fig. 1.2a, step 4).

Sandhu et al. formed 3C-SiC on SiO_2 films by anodically bonding 3C-SiC to aluminosilicate glass [69]. First, a 100-nm thick SiO_2 was created on a 3C-SiC/Si wafer by thermal oxidation at 1100 °C. Secondly, the SiO_2/SiC/Si structure was pressed onto an aluminosilicate glass plate and anodically bonded. Finally, the 3C-SiC and Si layers on the opposite side of the anodically bonded side was removed by RIE and wet etching, respectively, which left the 3C-SiC on glass structure as shown in Fig. 1.2b.

The wafer bonding technique combined with the smart cut technique was applied to create α-SiC on SiO_2/Si wafers [70, 71]. Figure 1.2c shows the process to produce single crystalline α-SiC. The process began with a commercial 6H-SiC wafer. Hydrogen protons were implanted into 6H-SiC, where the H^+ peak concentration was contributed at approximately 1.3 μm below the surface of SiC. This proton peak

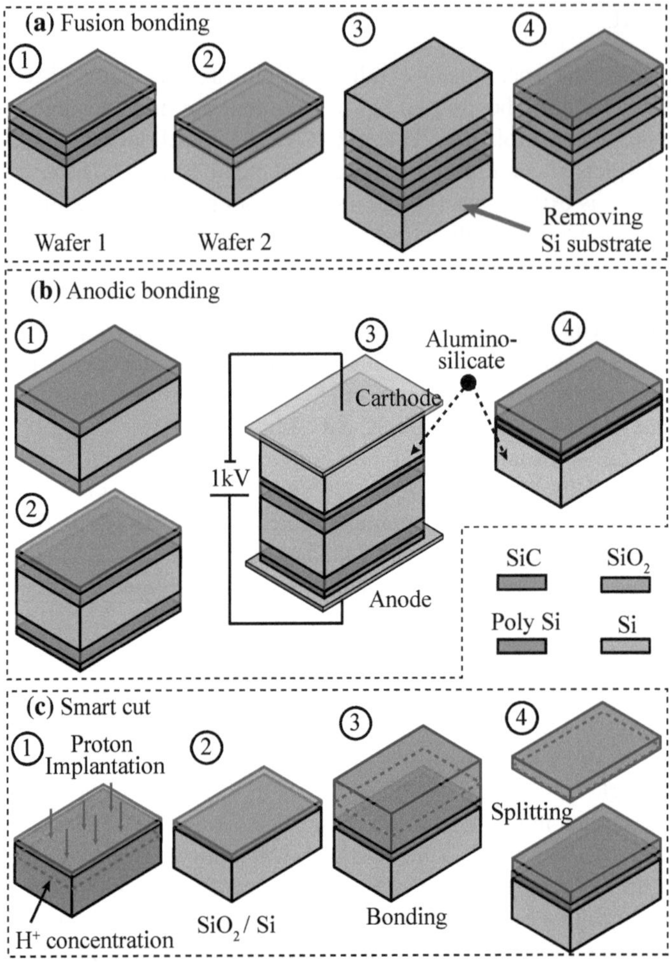

Fig. 1.2 Wafer bonding processes. **a** Fusion bonding. **b** Anodic bonding. **c** Smart cut process

concentration location can be controlled by the implant energy and is used to split the SiC wafer. Subsequently, the implanted SiC wafer was bonded onto a SiO_2/Si substrate. Finally, the SiC film was split along the peak concentration by a high temperature annealing step, forming a 1.3 μm thick single 6H-SiC layer over the carrier substrate.

Wafer bonding techniques demonstrated the possibility of making single crystalline 3C-SiC on insulators, offering more options to fabricate complicated and precise single crystalline structures [69, 72]. Additionally, the possibility of producing α-SiC on Si wafers can reduce the wafer cost and also made the fabrication of α-SiC films possible with conventional micromachining. In SiC based MEMS devices, common structures include the suspended SiC in the form of SiC diaphragms, double-

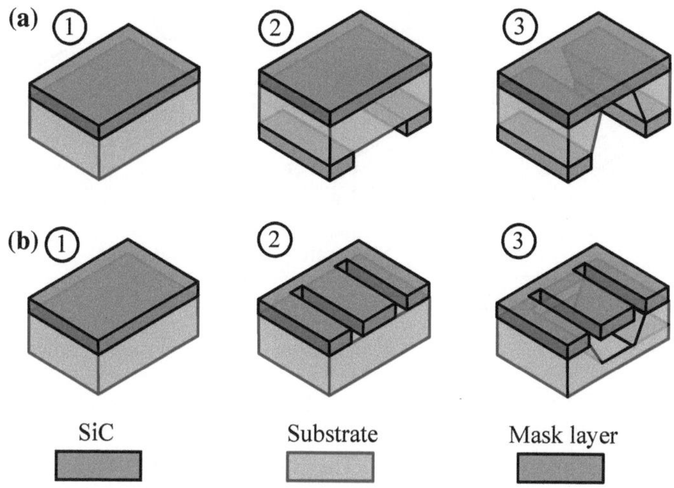

Fig. 1.3 Bulk patterning process of SiC, which releases SiC structure from the substrate. **a** The substrate is patterned from back side. **b** The substrate is etched from front side

side clamped beams, and cantilevers. To obtain these structures, SiC is patterned by employing the bulk process or surface process which are presented as follows.

Bulk Process

In bulk micro-machining, SiC is released by selectively removing the substrate underneath. This process is very important for bulk SiC wafers and SiC on Si wafers, where a sacrificial layer is not available [73].

Figure 1.3 presents the general concept of a bulk process which is applied to a 3C-SiC/Si wafer. In this process, SiC is etched by plasma or wet etching as mentioned above. Noting that SiC etching rate is significantly slower than the Si, and that SiC is inert to various Si etchants, such as TMAH and low-temperature KOH, the underlying Si layer can be patterned without damaging the SiC layer.

Releasing SiC structures from bulk α-SiC wafers (e.g. 4H-SiC and 6H-SiC bulk wafers) is more complicated as it takes a long time to remove a thick SiC layer. Therefore, wafer milling and laser machining are alternative methods, that are widely employed to accelerate the removal of SiC [74].

Surface Process

Surface patterning process can be applied to SiC wafers having a sacrificial layer such as SiO_2 and Si_3N_4 [75]. The suspended SiC structure is released by selectively removing the sacrificial layer. Figure 1.4 shows a general surface process applied to SiC on insulator wafers to form a free standing SiC cantilever. In this process, the first step is patterning the top SiC layer, and then the sacrificial layer is etched to

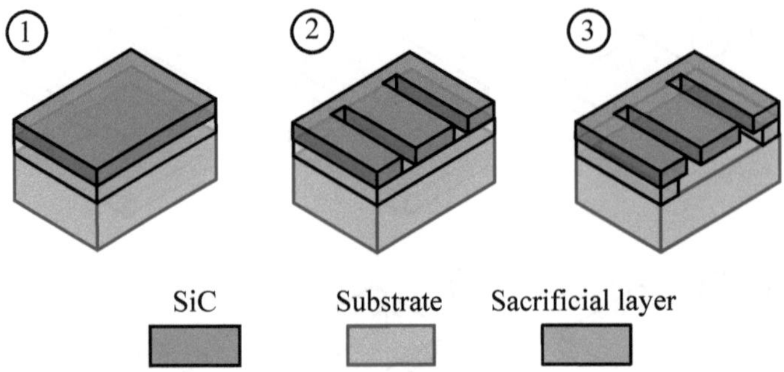

Fig. 1.4 Surface process of SiC where a sacrificial layer is removed by under etching from the top surface

release a SiC structure from substrates. The surface process is mainly deployed for polycrystalline, nanocrystalline, and amorphous SiC, as they can be deposited onto an insulator substrate (SiCOI wafer).

1.3 Piezoresistive Effect of Silicon Carbide

This section presents the definition of the piezoresistance and the gauge factors measured in different SiC films as well as nanowires. Typical SiC piezoresistive sensors are also described.

1.3.1 Definition of the Piezoresistance

For given a resistor with its resistivity, length, width and thickness being denoted as ρ, l, w and t, respectively, the resistance is $R = \rho l/(wt)$. When applying a strain, the dimensions and the resistivity of the resistor will be changed, leading to a change in the resistance:

$$\frac{\Delta R}{R} = \frac{\Delta\rho}{\rho} + \frac{\Delta l}{l} - \frac{\Delta w}{w} - \frac{\Delta t}{t} = \frac{\Delta\rho}{\rho} + (1 + 2\gamma)\varepsilon \qquad (1.1)$$

where $\varepsilon = \Delta l/l$ is the strain in the longitudinal direction while γ is the Poisson's ratio of the material. The piezoresistance is also quantified using a parameter named the gauge factor, which is the relative resistance change per unit strain:

$$GF = \frac{\Delta R/R}{\varepsilon} = \frac{\Delta\rho/\rho}{\varepsilon} + (1 + 2\gamma) \qquad (1.2)$$

For mechanical sensing application, materials with large gauge factors are desired to make sensing elements. In the case of metals, the resistivity is almost unchanged under stress, thus the gauge factor depends mainly on the change in resistors' geometry, which is typically below 2 [76, 77]. However, in semiconductors, applying strain could lead to a significant change in the resistivity due to energy band modification, thus results in much higher gauge factor than metals [1].

A large number of theoretical studies have been conducted to elucidate the piezoresistive phenomenon in semiconductors following the experimental work of Smith [78–80]. The physical mechanisms of the piezoresistance have been developed based on Bardeen and Shockley's model, which explains the change in carrier mobility in a silicon subjected to a deformation potentials [81–86]. Accordingly, the energy bands of semiconductors is modified due to strain, leading to the redistribution of carriers in these bands. This results in a change in the effective mass, and mobility of charge carriers, and thus the conductivity of materials. The piezoresistive effect in SiC was then theoretically explained based on these models in which the piezoresistance in n-type SiC is caused by the modification of the bottom conduction bands, while p-type is dominated by the change in energy structure of the top valence bands.

1.3.2 *Piezoresistance of Silicon Carbide*

Figure 1.5 illustrates the two typical methods utilized to experimentally characterize the piezoresistance of SiC [14, 74, 77]. In the bending beam technique, a piezoresistive layer is formed at the fixed end of the beam, and a uniaxial strain is applied by deflecting the other free end. The experimental approach employing diaphragm, on the other hand, will induce a biaxial strain to the piezoresistive element, when the diaphragm is subjected to a pressure.

The piezoresistance in SiC were preliminary reported in the early 1970s [87–89]. Nevertheless, because the wafers employed in these experimental work were involved to the Lely process, which was very irregular, these results were not widely accepted [90]. The first systematic study on the piezoresistance of high quality SiC was not reported until the 90s.

Table 1.3 presents a summary of experimental work on SiC piezoresistance. Accordingly, these studies were conducted in different poly types such as single α–SiC, β–SiC, poly crystals, and amorphous SiC. In addition, the piezoresistance was investigated in n-type and p-type with carrier concentrations raging from low dope to high dope. The absolute gauge factor ($|GF|$) of single crystal SiC was found to be approximately $20 \sim 30$ at low temperature, and decreased to nearly 10–15 at above 300 °C. On the other hand, poly SiC had relatively small gauge factor in comparison to single crystal SiC.

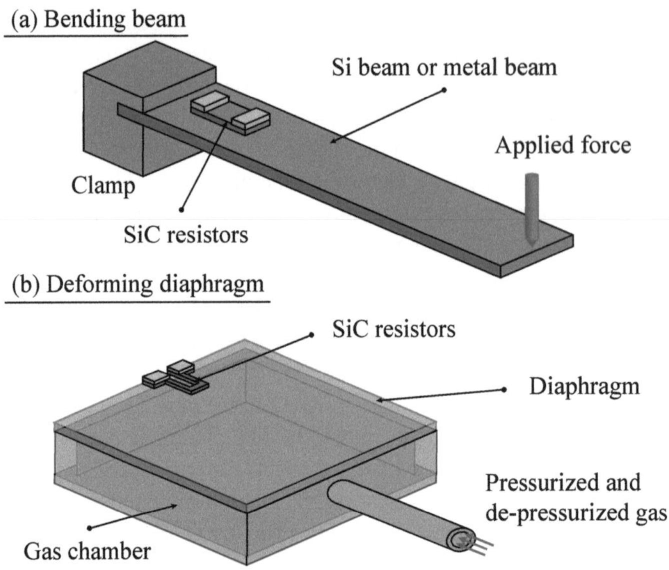

Fig. 1.5 Typical setup used to investigate piezoresistance: **a** Bending beam; and **b** Deforming diaphragm

The Piezoresistance of Single Crystal 3C-SiC

Studies on the piezoresistance of 3C-SiC has attracted more interest than other poly types thanks to its capability of deposition on Si [90, 91]. Shor et al. reported one of the first work on high quality n-type single crystal 3C-SiC APCVD-deposited on Si [90]. The gauge factor of the unintentionally doped 3C-SiC film with a carrier concentration of approximately 10^{16}–10^{17} cm^{-3} was found to be −31.8 at room temperature. Following this work, numerous studies were carried out on 3C-SiC films which were deposited on Si utilizing different growth techniques [91–94]. For instance, employing hot mesh CVD (HMCVD) to reduce the deposition temperature bellow 1000 °C, Yasui et al. found a gauge factor of −27 in the n-type 3C-SiC [59]. In addition, another work on selectively deposited n-type SiC found a gauge factor of −18 [91]. To eliminate a large leakage current from SiC to the Si substrate, 3C-SiC was transferred or deposited onto insulating substrates. As such, employing fusion bonding, Wu et al. obtained a gauge factor of −18 in an unintentionally doped n-type 3C-SiC film transferred onto SiO_2/Si [95]. In addition, the gauge factor of n-type SiCOI wafers developed using a bonding-free technique was reported to be −17.8 [96].

The piezoresistance in n-type 3C-SiC were also investigated at elevated temperatures, since it is mainly aimed for sensing devices in harsh environments. Figure 1.6 represents the variation of the piezoresistance in n-type single 3C-SiC films at different temperatures. Evidently, in low dope samples, the gauge factor decreased by approximately 50% when temperatures reaching above 400 °C in comparison to that at room temperature. Nevertheless, in high doped samples, the gauge factor was relatively stable when increasing temperature up to 450 °C.

Table 1.3 The gauge factors (GF) of SiC reported in the literature

Polytype	Growing process	Type	Dopant	Carrier concentration (cm^{-3})	GF Room temp	GF High temp	Orientation	Stress type
Single 3C-SiC [90]	APCVD	n	Nitrogen	$\sim 10^{18}$	−31.8	−18 (450 °C)	[100]	Uniaxial
Single 3C-SiC [59]	HMCVD	n	Nitrogen	$\sim 10^{18}$	−27	–	[100]	Uniaxial
Single 3C-SiC [95]	APCVD	n	Nitrogen	Unintentionally doped	−18	−7 (400 °C)	[100]	Biaxial
Single 3C-SiC [94]	LPCVD	n	Nitrogen	$0.4 \sim 2 \times 10^{17}$	−24.8	−11 (450 °C)	[100]	Biaxial
Single 3C-SiC [107]	APCVD	n	Nitrogen	Highly doped	−16	−12.5 (400 °C)	[100]	Uniaxial
Single 4H-SiC [105]	–	n	Nitrogen	1.5×10^{19}	20.8	–	(0001) plane	Uniaxial
Single 6H-SiC [99]	–	n	Nitrogen	3.8×10^{18}	−29.4	−17 (250 °C)	(0001) plane	Uniaxial
Single 6H-SiC [100]	–	n	Nitrogen	2×10^{19}	−22	−11 (250 °C)	(0001) plane	Uniaxial
Single 6H-SiC [100]	–	p	Aluminum	2×10^{19}	27	12 (250 °C)	(0001) plane	Uniaxial
Poly 3C-SiC [108]	LPCVD	n	Nitrogen	Low doped	−10	–	–	Biaxial
Poly 3C-SiC [95]	LPCVD	n	Nitrogen	Unintentionally doped	−2.1	–	–	Biaxial
Poly 3C-SiC [106]	LPCVD	p	Boron	$10^{18} \sim 10^{20}$	10	7 (200 °C)	–	Uniaxial
Amorphous SiC [112, 113]	PECVD	n	Nitrogen	–	49	–	–	Uniaxial
	Sputtering	n	Nitrogen	–	31	–	–	Uniaxial

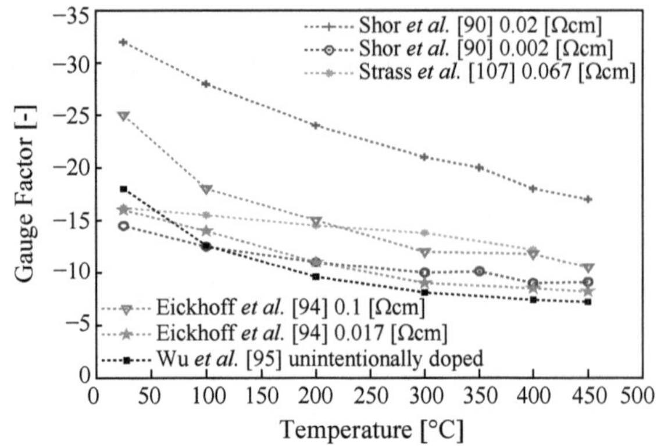

Fig. 1.6 The piezoresistance of n-type 3C-SiC at elevated temperatures. The diminution of the gauge factor with increasing temperature is more significant at low temperature and tends to stabilize at higher temperatures

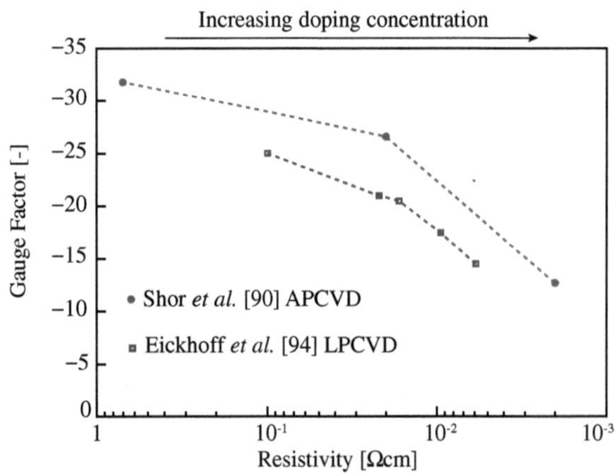

Fig. 1.7 The carrier concentration (or doping-level) dependence of the piezoresistance in n-type 3C-SiC

A number of works were conducted to investigate the dependence of 3C-SiC piezoresistance on carrier concentration which are plotted in Fig. 1.7. Evidently, a decrease in gauge factors was observed when increasing carrier concentrations. Additionally, the experimental data also indicated that the piezoresistance of high dope n-type 3C-SiC films was relatively stable at high temperatures compared to the low dope samples. As such, at 400 °C, gauge factors dropped by approximately ∼20 and ∼40% for high dope and low dope, respectively.

To investigate the dependence of gauge factors on crystal directions, Shor et al. fabricated n-type 3C-SiC piezoresistive elements aligned in different orientations such as [100] and [110]. The piezoresistive effect in these n-type samples showed a significant orientation dependence; as such longitudinal [100] direction exhibited a much larger gauge factor than longitudinal [110] and transverse [100]. This indicated that the crystal direction should be taken into account when designing SiC piezoresistive sensors.

The mechanism of the piezoresistance in n-type 3C-SiC was theoretically explained by Toriyama et al., in which the gauge factor of n-type was numerically estimated utilizing the deformation potential theory and electron transport phenomenon [97, 98]. Accordingly, at the bottom of the n-type 3C-SiC conduction bands in k-space, there are six energy valleys, locating in [100] equivalent orientation. When applying a compressive stress along [100] direction, the longitudinal valleys in [100] direction will shift downwards with respect to the other transverse valleys located in [010] and [001] axes. As a result, due to the change in the relative energy, electrons will move to lower energy levels. This means that more electrons will enter the valley with lower energy in [100] direction, and fewer electrons will populate in [010] and [001] valleys, as presented in Fig. 1.8a. Consequently, because the electron effective mass in longitudinal direction is greater than that of transverse direction, the redistri-

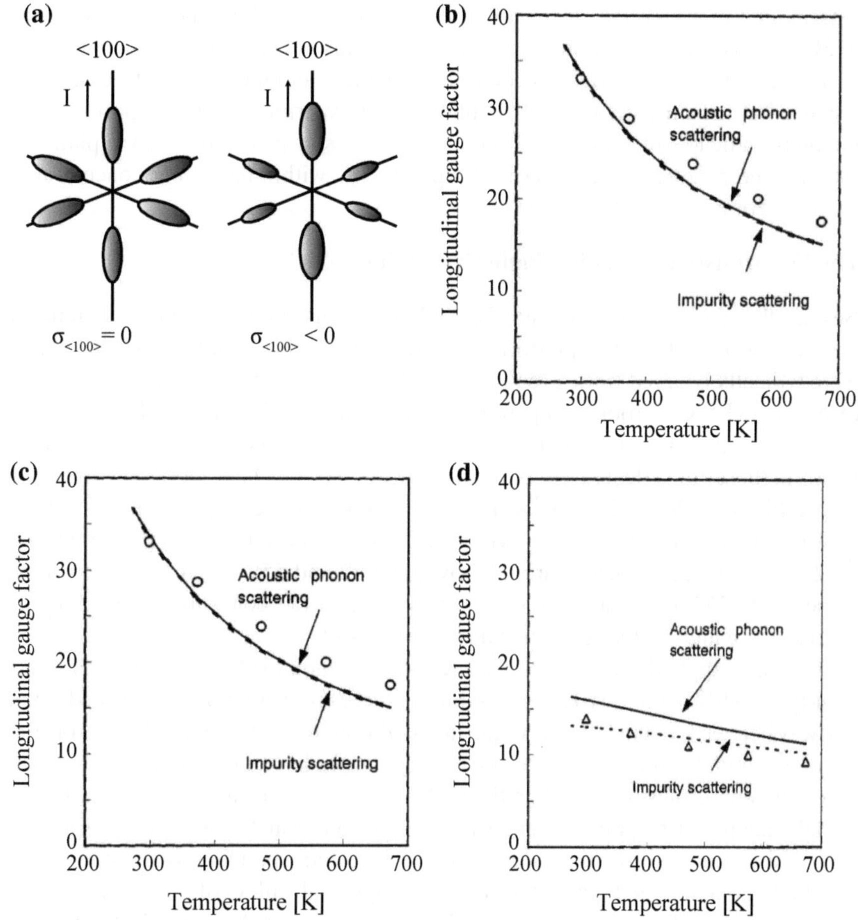

Fig. 1.8 **a** Illustration of the carrier re-population in the 6-valleys of 3C-SiC conduction band under a compressive stress in [100] orientation. **b, c, d** Theoretical estimation of the piezoresistance in n-type 3C-SiC. *Solid line*: acoustic phonon scattering; *Dart line*: impurity scattering; *Dot points*: experimental data for comparison with different carrier concentrations of **b** 10^{18} cm^{-3}; **c** 10^{19} cm^{-3}; **d** 10^{20} cm^{-3}, respectively; Reprint with permission from [98]. © American Institute of Physics

bution of electrons in n-type 3C-SiC due to the [110] compressive stress will cause its resistivity to increase. This explanation is in solid agreement with experimental data that n-type SiC exhibits negative gauge factor along [100] direction. In addition, the piezoresistance of n-type 3C-SiC can be quantified employing the following equation [97]:

$$G = 1 + 2\nu - \frac{\Xi_u(L-1)}{3k_B T(2L+1)}(2+\nu')\frac{F_{-1/2}}{F_{1/2}} \tag{1.3}$$

where Ξ_u is an independent constant of the deformation energy, F is the Fermi-Dirac integral, k_B is the Boltzmann constant, $L = \mu_l/\mu_t$ is the ratio between the longitudinal to transverse electron mobilities, and ν and ν' are the Poisson's ratios of 3C-SiC and the Si substrate, respectively. It is also evident that, the gauge factor of n-type depends on carrier concentrations and temperatures. In addition, the consistence between the theoretical calculation and experimental data reported in the literature indicates the validity of the electron transfer mechanisms to explain and estimate the piezoresistive effect in n-type 3C-SiC within the career concentration and temperature range mentioned above.

The Piezoresistive Effect in Single Crystalline α-SiC

Besides the work on single crystal 3C-SiC, numerous studies have been conducted on α-SiC where a bulk SiC platform was preferred. In those studies, 6H-SiC wafers were typically sourced from Cree Research Inc. [24]. The longitudinal gauge factor of n-type 6H-SiC at room temperature with doping level of 1.8×10^{17} cm^{-3} and 3.3×10^{18} cm^{-3} were reported to be −35 and −29.4, respectively [99]. The gauge factor in those films decreased by nearly 40% at 300 °C. Other works on high dope n-type 6H-SiC ($N_d \approx 2 \times 10^{19}$ cm^{-3}) and p-type 6H-SiC ($N_a \approx 2 \times 10^{19}$ cm^{-3}) showed gauge factors of 22 for n-type, and 27 for p-type at room temperature [100]. At 250 °C, the gauge factors dropped by about 52 and 55% for n-type and p-type, respectively. Figure 1.9 also shows that increasing carrier concentration leading to a decrease in gauge factor in n-type and p-type 6H-SiC.

Toriyama applied the electron transfer phenomenon explain the piezoresistance of n-type 6H-SiC [101, 102]. Accordingly, the energy minima are located about halfway between M and L points of the first Brillouin zone of n-type 6H-SiC, forming semi-ellipsoidal valleys. The constant energy surface has a dumbbell-like minimum when the energy level is smaller than 12 meV and it transforms into a full ellipsoidal centered at M point for energy levels greater than 12 meV [103]. Since the piezoresistance was investigated at above 298 K, corresponding to an energy level (3/2 $k_B T$) greater than 28 meV, the six semi-ellipsoidal valleys of n-type 6H-SiC can be simplified as three equivalent full ellipsoids located at M points [103, 104], as represented in Fig. 1.10a. Applying a tensile stress in [1100] orientation will cause these bands to deform, as a result unbalancing the symmetry of the three valleys, Fig. 1.10b Inset. Consequently, electrons will enter lower energy levels from higher energy valleys, which modifies the effective mass along with the mobility of carriers,

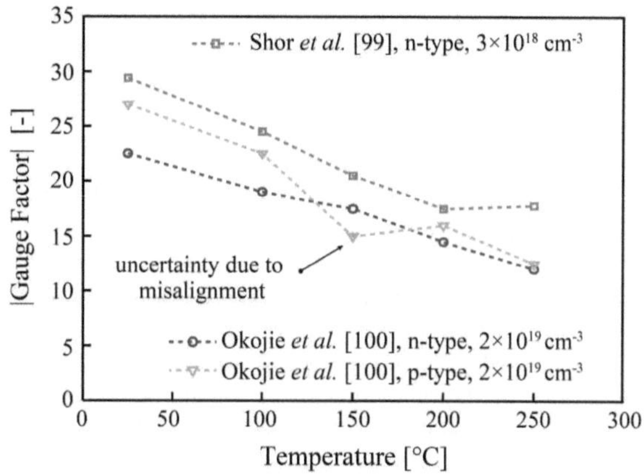

Fig. 1.9 The dependence of the piezoresistance of 6H-SiC on carrier concentration and temperature

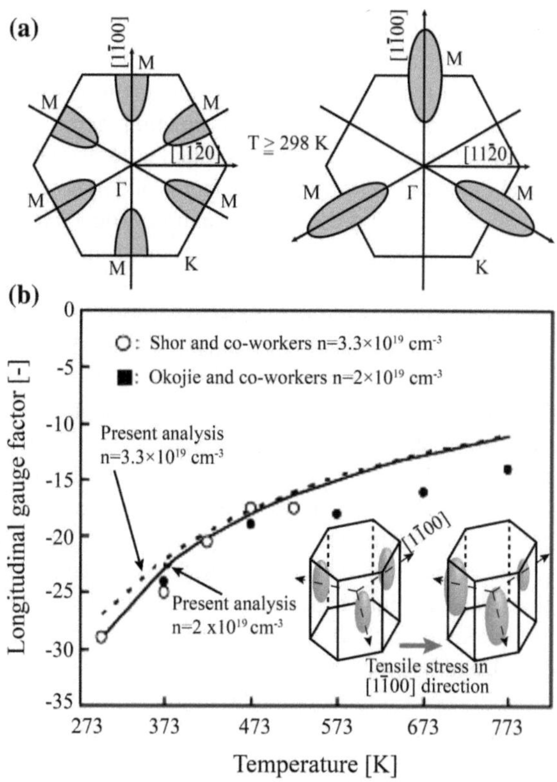

Fig. 1.10 **a** The conduction bands located at M points in the first Brillouin zone (six semi-ellipsoids and equivalent three full ellipsoids); **b** A comparison between numerical calculation and experimental data (*Inset* the deformation of the energy surfaces of three full valleys under a tensile stress $\sigma_{1\bar{1}00}$). Reprint with permission from [102]

and thus changing the resistivity of n-type 6H-SiC [102]. The gauge factor calculated using this model was in consistent with the experimental data reported in [99, 100].

The piezoresistance of another widely-used α-SiC, 4H-SiC, was reported in [105], where a n-type film with a thickness of 1 μm and a doping level of 1.5×10^{19} cm^{-3}

was epitaxially deposited on a bulk 4H-SiC bulk wafer. The transverse gauge factor of the n-type film was measured to be 20.8 at room temperature. The gauge factor in n-type 4H-SiC decreased at temperatures varying from 25 °C to 500 °C, and surprisingly increased when increasing temperature to 800° [138]. Although the mechanisms of this anomalous phenomenon has not been elucidated, the authors in [138] suggested that the packaging process could be a considerable factor leading to this unusual change in the gauge factor.

The Piezoresistive Effect in Poly, Amorphous, and Ceramic SiC

The piezoresistance of poly and amorphous SiC is also of interest, owing to their capability of deposition onto an insulator. P-type poly 3C-SiC with boron as the dopant was grown on SiC on SiO_2 employing plasma assisted CVD [106]. The gauge factor of this film with two impurity concentrations of 10^{18} and 10^{20} cm^{-3} were reported to be 7–10, respectively. These films also showed a relatively stable gauge factors up to 200 °C. Deploying LPCVD, Strass et al. deposited n-type textured and non-textured poly 3C-SiC films on a dioxide substrate [107]. An anisotropic property in the piezoresistance of the textured poly 3C-SiC was observed, showing gauge factors of 10 in [110] direction, and 6.1 in [100] direction, respectively. On the other hand, randomly oriented poly 3C-SiC did not show significant directional dependence with a gauge factor of 5 in both [100] and [110] directions. Poly 3C-SiC also exhibits doping concentration dependence and temperature dependence as single 3C-SiC [108]. For instance, the gauge factors at room temperature of poly SiC films with conductivities of 20 and 120 $S.cm^{-1}$ were measured to be −9.5 and −6, respectively. All of the above experimental data also indicated that poly 3C-SiC had smaller gauge factors than single 3C-SiC. The diminution of the piezoresistance in poly SiC could be caused by carrier scattering at grain boundaries and/or reasoned from the random alignment of crystal grains [109–111].

Piezoresistance of amorphous SiC has also been reported recently, in which the amorphous SiC films were deposited on a dioxide substrate employing CVD or sputtering [112, 113]. To form n-type amorphous SiC, nitrogen precursor was used in situ, showing longitudinal gauge factors ranging from 31 to 49.

Kishimoto et al. reported the piezoresistance of ceramic 6H-SiC with aluminum, boron, and gallium being used as dopants [114, 115]. Relatively large piezoresistive coefficients of 20×10^{-11}, 30×10^{-11} and 60×10^{-11} Pa^{-1} for Al-doped, Ga-doped, and B-doped samples, respectively. It should also be pointed out that, to date, there has been no report on the piezoresistance of amorphous and ceramic SiC at high temperature.

The Piezoresistance of Nanoscale SiC

Investigation into the piezoresistance of nanoscale SiC has also attracted significant interest owing to exciting physical properties as well as the capability of miniaturization at low dimension structures [116–118]. In addition, the extremely high gauge factor in Si nanowires reported by He and Yang have triggered the studies on the piezoresistance of nanowires and nano thin film SiC [76, 119–122].

Table 1.4 Piezoresistive effect of SiC nano structures

Authors	Poly type	Diameter [nm]	Gauge factor
Shao et al. [121]	3C-SiC	150	−6.9
Zeng et al. [122]	3C-SiC	320	14.1
Bi et al. [123]	3C-SiC	230	4.5–46.2
Gao et al. [124]	6H-SiC	170	25.6–79
Nakamura [126]	–	3 (thickness)	20–60

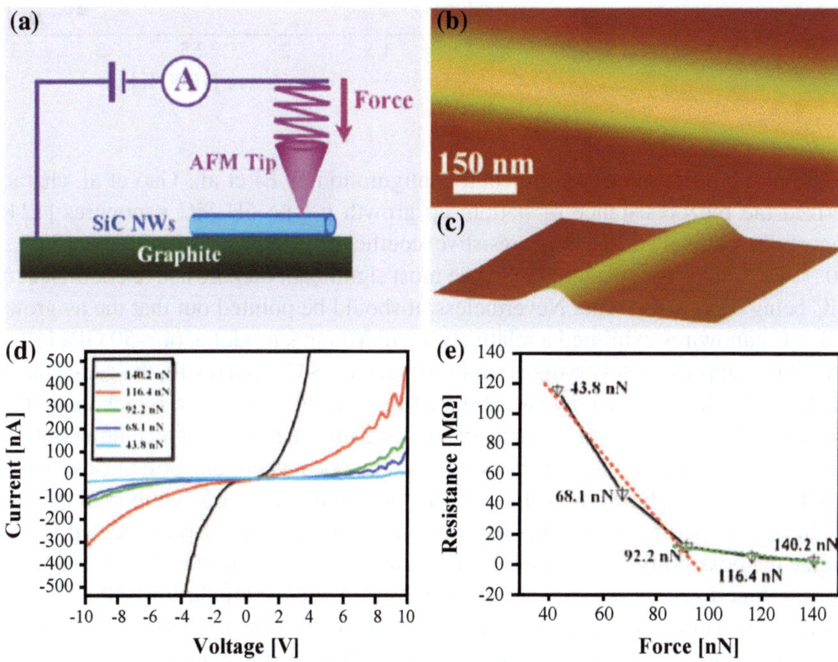

Fig. 1.11 **a** Experimental setup to induce local compressive strains into bottom up 6H-SiC nanowires; **b** and **c** AFM images of the as grown and transferred SiC nanowire; **d** the change of I–V curves under applied forces; **e** The resistance change of the nanowire under applied forces. Reprint with permission from [124]

By transferring bottom-up grown SiC nanowires onto nano actuators, Shao et al. [121] and Zeng et al. [122] developed a technique to measure the piezoresistance of the nanowires in scanning electron microscopy (SEM). The gauge factors of SiC nanowires reported on these works were measured to be −6.9 and 14.1, respectively, which were about 2–4 times smaller than bulk 3C-SiC (Table 1.4).

By pressing an AFM tip against a 3C-SiC nanowire placed on a graphite substrate, Bi et al. induced a locally concentrated strain into the nanowire [123]. The gauge factor of this top-down growth nanowire with respect to the local strain was found to be 4.5–46.2 (Fig. 1.11).

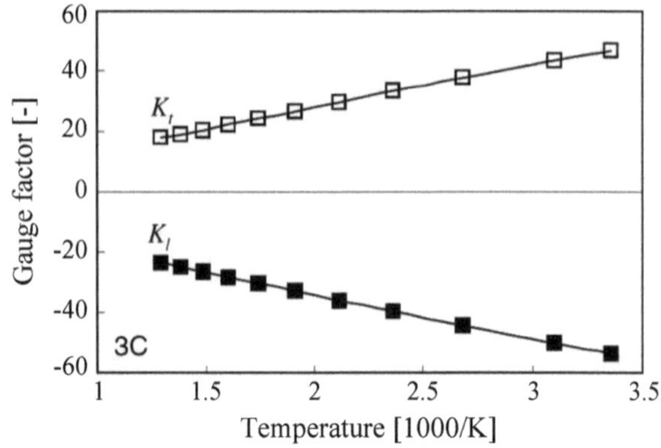

Fig. 1.12 Estimation of gauge factors in [100] direction of n-type 3C-SiC nano thin films with a carrier concentration of 10^{19} cm^{-3} employing the first principle approach. Reprint with permission from [126]

Employing the same experimental configuration as Bi et al., Gao et al. characterized the piezoresistance of bottom-up growth p-type 6H-SiC nanowires [124], reporting relatively large piezoresistive coefficient ranging from 51.2 to 159.5 $\times$ 10^{-11} Pa^{-1}. This is considered to be the most significant piezoresistive coefficient of SiC being reported to date. Nevertheless, it should be pointed out that the as-grown 6H-SiC nanowires exhibited a relatively small Young's modulus of $\sim$50 GPa [124], which was approximately 6 times smaller than bulk SiC. As a result, the gauge factor of these 6H-SiC nanowires was estimated to be from 25.6 to 79, which cannot be considered as giant piezoresistance. In addition, the measurement technique utilized in this work was different from conventional methods where uniform uni-axial or bi axial strains are induced, thus it is hard to compared to previously reported studies.

Besides experimental work, numerical calculations were also conducted to elucidate the possibility of the giant piezoresistance in nanoscale SiC [125]. Using the density functional theory, Nakamura et al. carried out the simulation of the piezoresistance of 3 nm thin n-type α and β-SiC(0001) nanosheets [126–128]. The first principle approach based simulation data indicated that the gauge factor of the SiC nanosheet at room temperature ranged from 30 to 60, and reduced by 50% at 500 °C (Fig. 1.12). These results suggested that the nano thin SiC films possessed a comparable piezoresistive effect to bulk SiC, and no giant piezoresistance was predicted based on the quantum confinement. It should also be pointed out that, in the theoretical analysis of Nakamura et al. the surface states (e.g. trapped charges) were not taken into account [129, 130]. Therefore, further studies need to be carried out to verify the existence of the giant piezoresistive effect in nano structured SiC.

1.3.3 Applications of Silicon Carbide Piezoresistive Effect

As well as the fundamental investigation into the piezoresistance of SiC, a large number of SiC piezoresistive MEMS sensors have been developed. Silicon carbide

Table 1.5 A list of pressure sensors employing of the piezoresistance of SiC

Authors	Poly type	Substrate	Sensitivity ([μV/V_{bias}]/kPa)		Released method
			RT	HT	
Shor et al. [90]	Single 3C-SiC	SiO_2/Steel	8.7	4.4 (350 °C)	–
Eickhoff et al. [91]	Single 3C-SiC	SOI	35	21 (200 °C)	ICP etching
Wu et al. [95]	Single 3C-SiC	SiO_2	25.8	9.2 (400 °C)	KOH
Ziermann et al. [131, 133]	Single 3C-SiC	SOI	20	11 (300 °C)	ICP etching
Berg et al. [132][a]	Single 3C-SiC	SOI	0.8	0.5 (300 °C)	–
Fraga et al. [112]	Amorphous SiC	SiO_2	48	–	KOH
Chung [134]	Poly 3C-SiC	SiO_2	1	0.25 (400 °C)	RIE
Wieczorek et al. [135][a]	6H-SiC	Bulk SiC	0.2	0.13 (400 °C)	Ultrasonic drilling
Okojie et al. [100, 136]	6H-SiC	Bulk SiC	1.2	0.6 (500 °C)	Electrochemical
Okojie et al. [137]	4H-SiC	Bulk SiC	1.74	0.7 (600 °C)	–
Okojie [138]	4H-SiC	Bulk SiC	2.9	3.4 (800 °C)	RIE
Akiyama et al. [74]	4H-SiC	Bulk SiC	2.6	–	Mechanical milling

[a]The sensitivity of the pressure sensors was calculated from the output voltage at a constant applied current

pressure sensors are among the most crucial applications employing the piezoresistance of this material, which are aimed to monitor the pressure level inside the combustion chambers of a engine at elevated temperatures [131–138]. A list of pressure sensors developed employing the piezoresistance in numerous SiC poly types is presented in Table 1.5. A general concept of these piezoresistive pressure sensors is shown in Fig. 1.13, in which different films such as 3C, 4H, 6H-SiC, and amorphous SiC were deployed to create the diaphragm structure of the devices. For example, in SiC on SiO_2/Si or SOI platform [69, 72, 139], the diaphragm was realized by etching the handling substrate employing either wet- (KOH) or dry- Si etching (RIE, ICP) [21, 140]. In addition, to create pressure sensors from a bulk SiC wafer, the SiC layer at the bottom can be patterned utilizing electrochemical etching or RIE. Nevertheless, due to the low etching rate of SiC (100 nm–1 µm/min [66, 67]), alternative technique such as milling or laser micro-machining were also applied to form SiC diaphragm [74, 135].

As shown in Table 1.5, the sensitivity of sensors significantly depends on the dimensions, as well as the Young's modulus of SiC diaphragms. In addition, increasing temperature leads to a decrease in the sensitivity of these sensors due to the

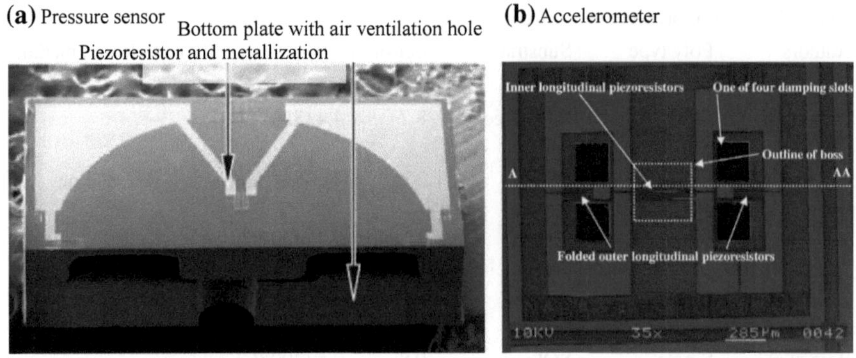

Fig. 1.13 Examples of SiC piezoresistive sensors: **a** Pressure sensors. **b** Accelerometers. Reprint with permission from [135, 141]

diminution of the piezoresistance in SiC at high temperatures. It should also be pointed out that most studies on the development of SiC pressure sensors were conducted as a proof of concept of using SiC piezoresistance for pressure monitoring at elevated temperatures. However, only a few work focused on the investigation of long term stability and reliability of these pressure sensors. For instance, the observation of the drift voltage of 4H-SiC pressure sensors working at 600 °C tested for 1000 h showed a profound agglomeration of a metal electrode (Au) and SiC [137]. In addition, the performance of 4H-SiC pressure sensors at 800 °C was also characterized, showing an anomalous property of the piezoresistance in SiC that the output of the sensors increased at temperatures above 400 °C [138]. The author in [138] suggested that further studies are required to elucidate this phenomenon.

Along with high temperature applications, the piezoresistance in SiC can also be utilized in high shock and high frequency devices. Figure 1.13b represents the concept of a SiC accelerometer which can measure extreme impact [141]. The accelerometers utilized piezoresistive 6H-SiC to sense an external mechanical impact, and a SiC proof mass located the centre of a diaphragm to convert the impact into an inertial force. In addition, by modifying the dimensions of the accelerometer, the first mode resonant frequency can be tuned from 200 to 800 kHz with a sensitivity varying from 40 to 250 nV/g. Furthermore, experimental data also showed that the as-fabricated sensor could detect impacts greater than 4×10^4 g, demonstrating the potential of piezoresistive effect in SiC for high shock and high frequency transducers.

The piezoresistance of SiC is also applicable for strain monitoring of hot sections inside combustion chambers. As such, Fraga et al. developed amorphous SiC strain sensors which exhibited a relative resistance change of 4.8% per 0.1% of strain at room temperature [142].

1.4 Summary of the Literature Review and the Aims of This Dissertation

1.4.1 Perspective of the Piezoresistive Effect in SiC

A large number of research has been carried out to investigate and elucidate the piezoresistive effect in SiC. Large gauge factor (approximately 30 at room temperature, about 10–18 at high temperature), demonstrated the potential of SiC as a mechanical sensing element.

The proof of concept of SiC pressure sensors used at temperatures above 500 °C has demonstrated the use of SiC piezoresistance under hostile conditions [132, 136, 138]. Additionally, thanks to its CMOS-compatibility, SiC piezoresistive sensors can be integrated into all-SiC based circuitry to replace Si in niche applications aiming at high power density and high temperatures [50]. Furthermore, owing to the high stiffness and excellent chemical inertness of SiC, the piezoresistive effect in SiC could be extended beyond electronic devices used at high temperatures to cover high-frequency [42] as well as bio applications [44]. As such, employing the large gauge factor in SiC could make the development of SiC based self-sensing resonators a promising possibility, and thereby can eliminate the need for optical measurements [143].

1.4.2 The Aims of This Research

From the literature, it is evident that among various poly types of SiC, there have been more studies on 3C-SiC grown on Si substrate than others. This is considerable due to the fact that 3C-SiC on Si could reduce the cost of SiC wafers and also be more compatible with MEMS processes. However, most of studies on single crystalline 3C-SiC reported in the literature were carried out on n-type semiconductor, while the piezoresistance of p-type single crystalline 3C-SiC have not been well understood. It is considered that the high temperature of approximately 1350 °C, required for the growth process, which is close to the melting point of Si is one of the main reason limiting the investigation into the piezoresistive effect in p type 3C-SiC. As such, the high growth temperature could accumulates the thermal mismatch between SiC and the Si substrate, and also affect the dopant redistribution in Si. To reduce the deposition temperature, great efforts have been made towards the growth process of cubic SiC [58, 59]. In the Queensland Micro and Nanotechnology Centre (QMNC), by utilizing the alternating-supply epitaxy (ASE), the growth temperature has been lowered to around 1000 °C [23].

It should also be noted that, for the case of silicon, numerous applications based on the piezoresistive effect have been developed using the p-type Si due to its large gauge factor. Therefore, the investigation into the piezoresistive effect in p-type 3C-SiC plays an important role in the development of 3C-SiC based sensors, as well as in the understanding of the influence of stress on general 3C-SiC electronic devices. Employing p-type single crystalline 3C-SiC films epitaxially deposited on large scale Si wafers at QMNC, this research aims to experimentally characterize and theoretically analyze the piezoresistance of p-type 3C-SiC for MEMS mechanical sensors. In particular, the following objectives are addressed in this book:

(I) Characterize the gauge factor of p-type single crystalline 3C-SiC. Based on the magnitude of the effect, the potential of the p-type 3C-SiC for MEMS mechanical sensors can be evaluated.

(II) Characterize the fundamental piezoresistive coefficients of the p-type single crystalline 3C-SiC.

(III) Explore the influence of scales by investigating the piezoresistive effect in SiC nano thin films.

(IV) Evaluate the influence of the temperature on the piezoresistive effect of p-type 3C-SiC.

(V) Investigate the piezoresistive effect in p-type 3C-SiC four-terminal resistors.

(VI) Compare the piezoresistive effect in p-type single crystalline SiC and nanocrystalline SiC grown on a Si substrate.

(VII) Explore the piezoresistance of top down fabricated SiC nanowires.

References

1. A.A. Barlian, W.T. Park, J.R. Mallon Jr., A.J. Rastegar, B.L. Pruitt, Review: semiconductor piezoresistance for microsystems. Proc. IEEE **97**(3), 513–552 (2009)
2. W.P. Eaton, J.H. Smith, Micromachined pressure sensors: review and recent developments. Smart Mater. Struct. **6**, 530–539 (1997)
3. S.S. Kumar, B.D. Pant, Design principles and considerations for the 'ideal' silicon piezoresistive pressure sensor: a focused review. Microsyst. Technol. **20**(7), 1213–1247 (2014)
4. M. Elwenspoek, H.V. Jansen, *Silicon Micromachining* (Cambridge University Press, London, 2004)
5. D.V. Dao, K. Nakamura, T.T. Bui, S. Sugiyama, Micro/nano-mechanical sensors and actuators based on SOI-MEMS technology. Adv. Nat. Sci: Nanosci. Nanotechnol. **1**(1), 013001 (2010)
6. S.M. Spearing, Materials issues in microelectromechanical systems (MEMS). Acta Materialia **48**(1), 179–196 (2000)
7. D.V. Dao, T. Toriyama, J. Wells, S. Sugiyama, Six-degree of freedom micro force-moment sensor for application in geophysics, in *2002 IEEE 15th International Conference on Micro Electro Mechanical Systems (MEMS)* (Las Vegas, USA, 2002), pp. 312–315
8. H. Yousef, M. Boukallel, K. Althoefer, Tactile sensing for dexterous in-hand manipulation in robotics-a review. Sens. Actuators A-Phys. **167**(2), 171–187 (2011)
9. M.-D. Nguyen, H.-P. Phan, K. Matsumoto, I. Shimoyama, A sensitive liquid-cantilever diaphragm for pressure sensor, in *2013 IEEE 26th International Conference on Micro Electro Mechanical Systems (MEMS)* (2013), pp. 617–620

10. N.C. Tsai, C.Y. Sue, Review of MEMS-based drug delivery and dosing systems. Sens. Actuators A-Phys. **134**(2), 555–564 (2007)
11. M. Li, H.X. Tang, M.L. Roukes, Ultra-sensitive NEMS-based cantilevers for sensing, scanned probe and very high-frequency applications. Nat. Nanotechnol. **2**(2), 114–120 (2007)
12. D.G. Senesky, B. Jamshidi, K.B. Cheng, A.P. Pisano, Harsh environment silicon carbide sensors for health and performance monitoring of aerospace systems: a review. IEEE Sens. J. **9**(11), 1472–1478 (2009)
13. C.V. Gungor, G.P. Hancke, Industrial wireless sensor networks: challenges, design principles, and technical approaches. IEEE Trans. Industr. Electron. **56**(10), 4258–4265 (2009)
14. M. Werner, G. Kroetz, H. Moller, M. Eickhoff, P. Gluche, M. Adamschik, C. Johnston, P.R. Chalker, High-temperature sensors based on SiC and diamond technology. Sens. Update **5**, 141–190 (1999)
15. G.H. Kroetz, M.H. Eickhoff, H. Moeller, Silicon compatible materials for harsh environment sensors. Sens. Actuators A-Phys. **74**, 182–189 (1999)
16. M. Willander, M. Friesel, Q.U. Wahab, B. Straumal, Silicon carbide and diamond for high temperature device applications. J. Mater. Sci. Mater. Electron. **17**(1), 1–25 (2006)
17. W.R. Fahrner, R. Job, M. Werner, Sensors and smart electronics in harsh environment applications. Microsyst. Technol. **7**, 138–1144 (2001)
18. P.G. Neudeck, R.S. Okojie, L.Y. Chen, High-temperature electronics - a role for wide bandgap semiconductors? Proc. IEEE **90**(6), 1065–1076 (2002)
19. M.R. Werner, W.R. Fahrner, Review on materials, microsensors, systems, and devices for high-temperature and harsh-environment applications. IEEE Trans. Industr. Electron. **48**(2), 249–257 (2001)
20. M. Mehregany, C.A. Zorman, N. Rajan, C.H. Wu, Silicon carbide MEMS for harsh environments. Proc. IEEE **86**(8), 1594–1610 (1998)
21. P.M. Sarro, Silicon carbide as a new MEMS technology. Sens. Actuators A-Phys. **82**, 210–218 (2000)
22. D. Masse, Market for GaN and SiC semiconductors set to rise 18x from 2012 to 2022. J. Microwave **56**(6), 55 (2013)
23. L. Wang, S. Dimitrijev, J. Han, A. Iacopi, L. Hold, P. Tanner, H.B. Harrison, Growth of 3C-SiC on 150-mm Si(100) substrates by alternating supply epitaxy at 1000°C. Thin Solid Films **519**, 6443–6446 (2011)
24. Commercial datasheet 2014; Durham, NC: Cree Inc., www.cree.com (2014)
25. R. Maboudian, C. Carraro, D.G. Senesky, C.S. Roper, Advances in silicon carbide science and technology at the micro-and nanoscales. J. Vac. Sci. Technol. A **31**(5), 050805 (2013)
26. W.D. Edwards, R.P. Beaulieu, Germanium piezoresistive element on a flexible substrate. J. Phys. E: Sci. Instrum. **2**(2), 613–615 (1969)
27. A.D. Bykhovski, V.V. Kaminski, M.S. Shur, Q.C. Chen, M.A. Khan, Piezoresistive effect in wurtzite n type GaN. Appl. Phys. Lett. **68**, 818 (1996)
28. V. Tilak, A. Vertiatchikh, J. Jiang, N. Reeves, S. Dasgupta, Piezoresistive and piezoelectric effects in GaN. Phys. Status Solidi (C) **6**, 2307–2311 (2006)
29. C.H. Park, B.H. Cheong, K.H. Lee, K.J. Chang, Structural and electronic properties of cubic, 2H, 4H and 6H SiC. Phys. Rev. B **49**(7), 4485–4493 (1994)
30. J.B. Casady, R.W. Johnson, Status of silicon carbide (SiC) as a wide-bandgap semiconductor for high-temperature applications: a review. Solid State Electron. **39**(10), 1409–1422 (1996)
31. V. Cimalla, J. Pezoldt, O. Ambacher, Group III nitride and SiC based MEMS and NEMS: materials properties, technology and applications. J. Phys. D: Appl. Phys. **379**(3–4), 149–255 (2003)
32. M. Wijesundara, R. Azevedo, *Silicon Carbide Microsystems for Harsh Environments* (Springer, Berlin, 2011)
33. Q. Zhang, R. Callanan, M.K. Das, S.H. Ryu, A.K. Agarwal, J.W. Palmour, SiC power devices for microgrids. IEEE. Trans. Power Electron. **25**(12), 2889–2896 (2010)
34. P. Friedrichs, SiC power devices as enabler for high power density-aspects and prospects. Mater. Sci. Forum **778**, 1104–1109 (2014)

35. G.S. Chung, J.M. Jeong, Fabrication of micro heaters on polycrystalline 3C-SiC suspended membranes for gas sensors and their characteristics. Microelectron. Eng. **87**, 2348–2352 (2010)
36. J.G. Lee, M.I. Lei, S.P. Lee, S. Rajgopal, M. Mehregany, Micro flow sensor using polycrystalline silicon carbide. J. Sens. Sci. Technol. **18**(2), 147–153 (2009)
37. T. Dinh, D.V. Dao, H.P. Phan, L. Wang, A. Qamar, N.T. Nguyen, P. Tanner, M. Rybachuk, Charge transport and activation energy of amorphous silicon carbide thin film on quartz at elevated temperature. Appl. Phys. Express **8**(6), 061303 (2015)
38. S. Ma, S. Wang, F. Iacopi, H. Huang, A resonant method for determining the residual stress and elastic modulus of a thin film. Appl. Phys. Lett. **103**(3), 031603 (2013)
39. G. Cheng, T.H. Chang, Q. Qin, H. Huang, Y. Zhu, Mechanical properties of silicon carbide nanowires: effect of size-dependent defect density. Nano Lett. **14**(2), 754–758 (2014)
40. S. Gong, N.K. Kuo, G. Piazza, GHz high-Q lateral overmoded bulk acoustic-wave resonators using epitaxial SiC thin film. J. Micromech. Syst. **21**(2), 253–255 (2012)
41. A.R. Kermany, G. Brawley, N. Mishra, E. Sheridan, W.P. Bowen, F. Iacopi, Microresonator with Q-factors over a million from highly stressed epitaxial silicon carbide on silicon. Appl. Phys. Lett. **104**, 801901 (2014)
42. Z. Wang, J. Lee, P.X.-L. Feng, Spatial mapping of multimode Brownian motions in high-frequency silicon carbide microdisk resonators. Nat. Commun. **5**, 5158 (2014)
43. C. Forster, V. Cimalla, V. Lebedev, J. Pezoldt, K. Brueckner, R. Stephan, M. Hein, E. Aperathitis, O. Ambacher, Group III-nitride and SiC based micro- and nanoelectromechanical resonators for sensor applications. Phisica Status Solidi A **203**(7), 1829–1833 (2006)
44. A. Oliveros, A. Guiseppi-Elie, S.E. Saddow, Silicon carbide: a versatile material for biosensor applications. Biomed. Microdevices **15**(2), 353 (2013)
45. S.E. Saddow, C.L. Frewin, C. Coletti, N. Schettini, E. Weeber, A. Oliveros, Single-crystal silicon carbide: a bio-compatible and hemocompatible semiconductor for advanced biomedical applications. Mater. Sci. Forum **679–680**, 824–830 (2011)
46. G. Gabriel, I. Erill, J. Caro, R. Gomez, D. Riera, R. Villa, Manufacturing and full characterization of silicon carbide based multi-sensor micro-probe for biomedical applications. Microelectron. J. **38**, 406–415 (2007)
47. S. Fujita, Wide-bandgap semiconductor materials: for their full bloom. Jpn. J. Appl. Phys. **54**(3), 030101 (2015)
48. H. Amano, Progress and prospect of the growth of wide-band-gap group III nitrides: development of the growth method for single-crystal bulk GaN. J. Jpn. Appl. Phys. **52**, 050001 (2013)
49. N.G. Wright, A.B. Horsfall, SiC sensors: a review. J. Phys. D: Appl. Phys. **40**, 6345–6354 (2007)
50. C.M. Zetterling, Integrated circuits in silicon carbide for high-temperature applications. MRS Bull. **40**(05), 431–438 (2015)
51. N.G. Wright, A.B. Horsfall, K. Vassilevski, Prospects for SiC electronics and sensors. Mater. Today **11**(1), 16–21 (2008)
52. J.A. Lely, Sublimation process for manufacturing silicon carbide crystals, United States Patent 2, 854, 364 (30.09.58) (1958)
53. Y.M. Tairov, V.F. Tsvetkov, Investigation of growth processes of ingots of silicon carbide single crystals. J. Cryst. Growth **43**, 209–212 (1978)
54. D. Chaussende, P.J. Wellmann, M. Pons, Status of SiC bulk growth processes. J. Phys. D: Appl. Phys. **40**, 6150–6158 (2007)
55. A. Itoh, H. Matsunami, Single crystal growth of SiC and electronic devices. Crit. Rev. Solid State Mater. Sci. **22**(2), 111–197 (1997)
56. M. Portail, M. Zielinski, T. Chasssagne, S. Roy, M. Nemoz, Comparative study of the role of the nucleation stage on the final crystalline quality of (111) and (100) silicon carbide films deposited on silicon substrates. J. Appl. Phys. **105**, 083505 (2009)
57. T. Fuyuki, T. Yoshinobu, H. Matsunami, Atomic layer epitaxy controlled by surface superstructure in SiC. Thin Solid Films **225**, 225 (1993)

58. H. Zhuang, L. Zhang, T. Staedler, X. Jiang, Low temperature hetero-epitaxial growth of 3C-SiC films on Si utilizing microwave plasma CVD. Chem. Vap. Depos. **19**, 29–37 (2013)
59. K. Yasui, H. Miura, M. Takata, T. Akadane, SiCOI structure fabricated by catalytic chemical vapor deposition. Thin Solid Film **516**, 644–647 (2008)
60. A. Gupta, C. Jacob, Selective epitaxy and lateral overgrowth of 3C-SiC on Si - a review. Prog. Cryst. Growth Charact. Mater. **51**, 43–69 (2005)
61. V. Heera, D. Panknin, W. Skorupa, P-type doping of SiC by high dose Al implatation-problems and progress. Appl. Surf. Sci. **184**, 307–316 (2001)
62. L. Wang, S. Dimitrijev, P. Tanner, J. Zou, Aluminum induced in situ crystallization of armourphous SiC. Appl. Phys. Lett. **94**, 181909 (2009)
63. S.A. Sakwe, R. Muller, P.J. Wellmann, Optimization of KOH etching parameters for quantitative defect recognition in n- and p-type doped SiC. J. Cryst. Growth **289**, 520–526 (2006)
64. D. Zhuang, J.H. Edgar, Wet etching of GaN, AlN, and SiC: a review. Mater. Sci. Eng. R-Rep. **48**(1), 1–46 (2005)
65. J.S. Shor, R.M. Osgood, A.D. Kurtz, Photoelectrochemical conductivity selective etch stops for SiC. Appl. Phys. Lett. **60**, 1001 (1992)
66. T.K. Hossain, S. MacLaren, J.M. Engel, C. Liu, I. Adesida, R. Okojie, The fabrication of suspended micromechanical structures from bulk 6H-SiC using an ICP-RIE system. J. Micromech. Microeng. **16**, 751–756 (2006)
67. S. Tanaka, K. Rajanna, T. Abe, M. Esashi, Deep reactive ion etching of silicon carbide. J. Vac. Sci. Technol. B. **19**(6), 2173–2176 (2001)
68. K.N. Vinod, C.A. Zorman, A.A. Yasseen, M. Mehregany, Fabrication of low defect density 3C-SiC on SiO_2 structures using wafer bonding techniques. J. Electron. Mater. **27**(3) (1998)
69. A. Sandhu, S. Jinno, Piezoresistive properties of 3C-SiC films anodically bonded to aluminosilicate glass substrates. Electron. Lett. **36**(6), 497–498 (2000)
70. P. Cong, D.J. Young, Single crystal 6H-SiC MEMS fabrication based on smart-cut technique. J. Micromech. Microeng. **15**, 2243–2248 (2005)
71. R. Yang, Z. Wang, J. Lee, K. Ladhane, D.J. Young, P.X.-L. Feng, 6H-SiC microdisk torsional resonators in a smart-cut technology. Appl. Phys. Lett. **104**, 091906 (2014)
72. W. Reichert, E. Obermeier, J. Stoemenos, β-SiC films on SOI substrates for high temperature applications. Diam. Relat. Mater. **6**(10), 1448–1450 (1997)
73. C.A. Zorman, R.J. Parro, Micro- and nanomechanical structures for silicon carbide MEMS and NEMS. Phys. Status Solidi B **245**, 1404–1424 (2008)
74. T. Akiyama, D. Briand, N.F. de Rooij, Piezoresistive n-type 4H-SiC pressure sensor with membrane formed by mechanical milling, in *Proceedings of the IEEE Sensors Conference 2011* (2011), p. 222
75. A.A. Yasseen, C.H. Wu, C.A. Zorman, M. Mehregany, Fabrication and testing of surface micromachined silicon carbide micromotors, in *Proceedings of the 12th IEEE International Conference on Micro Electro Mechanical Systems* (Florida, USA, 1999), pp. 644–649
76. A.C.H. Rowe, Piezoresistance in silicon and its nanostructures. Mater. Res. Soc. **29**(6), 731–744 (2014)
77. J.C. Doll, B.L. Pruitt, *Piezoresistor Design and Applications* (Springer, Berlin, 2013)
78. G.L. Bir, G.E. Pirkus, *Symmetry and Strain-Induced Effects in Semiconductors* (Wiley, New York, 1974)
79. Y. Sun, S.E. Thompson, T. Nishida, *Strain Effect in Semiconductor: Theory and Device Applications*, 1st edn. (Springer, Berlin, 2009)
80. C.S. Smith, Piezoresistance effect in germanium and silicon. Phys. Rev. **94**, 42 (1954)
81. J. Bardeen, W. Shockley, Deformation potentials and mobilities in non-polar crystals. Phys. Rev. **80**, 72–80 (1950)
82. C. Herring, Transport properties of a many valley semiconductor. Bell Syst. Tech. J. **34**, 237–290 (1955)
83. C. Herring, E. Vogt, Transport and deformation potential theory for many-valley semiconductors with anisotropic scattering. Phys. Rev. **101**, 944 (1956)
84. D. Long, Stress dependence of the piezoresistance effect. J. App. Phys. **32**, 2050–2051 (1961)

85. Y. Kanda, The piezoresistive effect effect of silicon. Sens. Actuators A-Phys. **28**, 83–91 (1991)
86. Y. Kanda, A graphical representation of the piezoresistance coefficients in silicon. IEEE Trans. Electron Devices **29**(1), 64–70 (1982)
87. I.V. Rapatskaya, G.E. Rudashevskii, M.G. Kasaganova, M.I. Iglitsin, M.B. Reifman, E.F. Fedetova, The piezoresistive effect coefficients of n-type α-SiC. Sov. Phys. Solid State **9**(12), 2833 (1968)
88. G.N. Guk, N.Y. Usoltseva, V.S. Shadrin, N.K. Prokopeva, The piezoresistive effect of cubic silicon carbide under hydrostatic compression. Sov. Phys. Semicond. **10**(1), 83 (1976)
89. G.N. Guk, V.M. Lyubimskii, E.P. Gofman, V.B. Zinovev, E.A. Chalyi, Temperature dependence of the piezoresistive effect constant π_{11} of n-type SiC(6H). Sov. Phys. Semicond. **9**, 104 (1974)
90. J.S. Shor, D. Goldstein, A.D. Kurtz, Characterization of n-type β-Sic as a piezoresistor. IEEE Trans. Electron Devices **40**(6), 1093–1099 (1993)
91. M. Eickhoff, H. Moller, G. Kroetz, J.V. Berg, R. Ziermann, A high temperature pressure sensor prepared by selective deposition of cubic silicon carbide on SOI substrates. Sens. Actuators A-Phys. **74**, 56–59 (1999)
92. A. Qamar, H.-P. Phan, D.V. Dao, P. Tanner, T. Dinh, L. Wang, S. Dimitrijev, The dependence of offset voltage in p-type 3C-SiC van der pauw device on applied strain. IEEE Electron Devices Lett. **36**(7), 708–710 (2015)
93. M. Eickhoff, H. Moller, J. Stoemenos, S. Zappe, G. Kroetz, M. Stutzmann, Influence of crystal quality on the electronic properties of n-type 3C-SiC grown by low temperature low pressure chemical vapor deposition. J. Appl. Phys. **95**, 7908 (2004)
94. M. Eickhoff, M. Stutzmann, Influence of crystal defects on the piezoresistive properties of 3C-SiC. J. Appl. Phys. **96**, 2878 (2004)
95. C.H. Wu, C.A. Zorman, M. Mehregany, Fabrication and testing of bulk micromachined silicon carbide piezoresistive pressure Sensors for high temperature applications. IEEE Sens. J. **6**(2), 316–324 (2006)
96. H.I. Kuo, C.A. Zorman, M. Mehregany, Fabrication and testing of single crystalline 3C-SiC devices using a novel SiC on insulator substrate, in *Proceedings of the Transducer 03 International Conference on Solid State Sensors and Actuators* (Boston, USA, 2003), pp. 724–745
97. T. Toriyama, S. Sugiyama, Analysis of the piezoresistive effect in n-type β-SiC based on electron transport and deformation potential theory, *International Symposium on Micromechatronic and Human Science* (2000), pp. 175–180
98. T. Toriyama, S. Sugiyama, Analysis of the piezoresistive effect in n-type β-SiC for high-temperature mechanical sensors. Appl. Phys. Lett. **81**, 2797 (2002)
99. J.S. Shor, L. Bemis, A.D. Kurtz, Characterization of monolithic n-type 6H-SiC piezoresistive sensing elements. IEEE Trans. Electron Devices **41**(5), 661–665 (1994)
100. R.S. Okojie, A.A. Ned, A.D. Kurtz, W.N. Carr, Characterization of highly doped n- and p-type 6H-SiC piezoresistors. IEEE Trans. Electron Devices **45**(4), 785–790 (1998)
101. T. Toriyama, S. Sugiyama, Analysis of piezoresistance in n-type 6H-SiC for high temperature mechanical sensors, *Proceedings of the Transducers 2003 International Conference on Solid State Sensors and Actuators* (Boston, USA, 2003), pp. 758–761
102. T. Toriyama, Piezoresistance consideration on n-type 6H SiC for MEMS-based piezoresistance sensors. J. Micromech. Microeng. **14**, 1445–1448 (2004)
103. T.K. Kinoshita, M. Itoh, M. Schadt, G. Pensl, Theory of the electron mobility in n-type 6H-SiC. J. Appl. Phys. **85**(12), 8193–8198 (1999)
104. W.J. Choyke, H. Matsunami, *Silicon Carbide: Recent Major Advances* (Springer Science & Business Media, Berlin, 2004)
105. T. Akiyama, D. Briand, N.F. de Rooiji, Design-dependent gauge factors of highly doped n-type 4H-SiC piezoresistors. J. Micromech. Microeng. **22**, 085034 (2012)
106. T. Homma, K. Kamimura, H.Y. Cai, Y. Onuma, Preparation of polycrystalline SiC films for sensors used at high temperature. Sens. Sensors A-Phys. **40**, 93–96 (1994)

107. J. Strass, M. Eickhoff, The influence of crystal quality on the piezoresistive effect of p-SiC between RT and 450°C measured by using microstructures, *Proceedings of the Transducer 97 International Conference on Solid State Sensors and Actuators* (Chicago, USA, 1997), pp. 1439–1442
108. M. Eickhoff, M. Moller, G. Kroetz, M. Stutzmann, Piezoresistive properties of single crystalline, polycrystalline, and nanocrystalline n-type 3C-SiC. J. Appl. Phys. **96**, 2872–2877 (2004)
109. J.Y.W. Seto, Piezoresistive properties of polycrystalline silicon. J. Appl. Phys. **47**, 4780 (1976)
110. X. Liu, C. Shi, R. Chuai, Polycrystalline silicon piezoresistive nano thin film technology, Solid State Circuits Technologies (2010)
111. V. Mosser, J. Suski, J. Goss, E. Obermeier, Piezoresistive pressure sensors based on polycrystalline silicon. Sens. Actuators A **28**(2), 113–132 (1991)
112. M.A. Fraga, M. Massi, H. Furlan, I.C. Oliveira, L.A. Rasia, C.F.R. Mateus, Preliminary evaluation of the influence of the temperature on the performance of a piezoresistive pressure sensor based on a-SiC film. Microsys. Technol. **17**, 477–480 (2011)
113. M.A. Fraga, H. Furlan, R.S. Pessoa, L.A. Rasia, C.F.R. Mateus, Studies on SiC, DLC and TiO2 thin films as piezoresistive sensor materials for high temperature application. Microsys. Technol. **18**, 1027–1033 (2012)
114. A. Kishimoto, D. Mutaguchi, H. Hayashi, Y. Numata, High temperature the piezoresistive effect properties of 6H-SiC ceramics doped with trivalent elements. Mater. Sci. Eng. B **135**, 145–149 (2006)
115. A. Kishimoto, Y. Okada, H. Hayashi, Improvement of the piezoresistive effect properties of silicon carbide ceramics through co-doping of aluminum nitride and nitrogen. Ceram. Int. **34**, 845–848 (2008)
116. K. Zekentes, K. Rogdakis, SiC nanowires: material and devices. J. Phys. D: Appl. Phys. **44**, 133011 (2011)
117. R. Wu, K. Zhou, C.Y. Yue, J. Wei, Y. Pan, Recent progress in synthesis, properties and potential applications of SiC nanomaterials. Prog. Mater. Sci. **72**, 1–60 (2015)
118. A. Lugstein, M. Steinmair, A. Steiger, H. Kosina, E. Bertagnolli, Anomalous piezoresistance effect in ultrastrained silicon nanowires. Nano Lett. **10**, 3204–3208 (2010)
119. R. He, P. Yang, Giant the piezoresistive effect effect in silicon nanowires. Nature Nanotechnol. **1**, 42–46 (2006)
120. A.C.H. Rowe, Silicon nanowires feel the pinch. Nature Nanotechnol. **3**(6), 311–312 (2008)
121. R. Shao, K. Zheng, Y. Zhang, Y. Li, Z. Zhang, The piezoresistive effect behaviors of ultra-strained SiC nanowires. Appl. Phys. Lett. **101**, 233109 (2012)
122. H. Zeng, T. Li, M. Bartenwerfer, S. Fatikow, Y. Wang, In situ SEM electromechanical characterization of nanowire using an electrostatic tensile device. J. Phys. D: Appl. Phys. **46**, 305501 (2013)
123. J. Bi, G. Wei, L. Wang, F. Gao, J. Zheng, B. Tang, W. Yang, Highly sensitive piezoresistance behaviors of n-type 3C-SiC nanowires. J. Mater. Chem. C **1**, 4514 (2013)
124. F. Gao, J. Zheng, M. Wang, G. Wei, W. Yang, Piezoresistance behaviors of p-type 6H-SiC nanowires. Chem. Commun. **47**, 11993–11995 (2011)
125. K. Nakamura, T. Toriyama, S. Sugiyama, Analysis on piezoresistive property of silicon carbide on the basis of first-principles calculation, *Proceedings of the 27th Sensor Symposium* (2010), pp. 1–16
126. K. Nakamura, T. Toriyama, S. Sugiyama, First-principles simulation on piezoresistivity in alpha and beta silicon carbide nanosheets. Jpn. J. Appl. Phys. **50**, 06GE05 (2011)
127. K. Nakamura, Y. Isono, T. Toriyama, First-principles study on piezoresistance effect in silicon nanowires. Jpn. J. Appl. Phys. **47**, 5132–5138 (2008)
128. K. Nakamura, Y. Isono, T. Toriyama, S. Sugiyama, Simulation of piezoresistivity in n-type single-crystal silicon on the basis of the first-principles band structure. Phys. Rev. B **80**, 045205 (2009)
129. Y. Yang, X. Li, Giant piezoresistance of p-type nano-thick silicon induced by interface electron trapping instead of 2D quantum confinement. Nanotechnology **22**, 015501 (2011)

130. L.M. Terman, An investigation of surface states at silicon/silicon oxide interface employing metal-oxide-silicon diodes. Solid State Electron. **5**, 285 (1962)
131. R. Ziermann, J.V. Berg, W. Reichert, E. Obermeier, M. Eichkoff, G. Kroetz, A high temperature pressure sensor with β-SiC piezoresistors on SOI substrates, *Proceedings of the Transducers 97 International Conference on Solid State Sensors and Actuators* (Chicago, USA, 1997), pp. 1411–1414
132. J.V. Berg, R. Ziermann, W. Reichert, E. Obermeier, Measurement of the cylinder pressure in combustion engines with a piezoresistive β-SiC-on-SOI pressure sensor, *Proceedings of the High Temperature Electronics Conference* (1998), pp. 245–249
133. R. Ziermann, J.V. Berg, E. Obermeier, F. Wischmeyer, E. Niemann, H. Moller, M. Eickhoff, G. Kroetz, High temperature piezoresistive β-SiC-on-SOI pressure sensor with on chip SiC thermistor. Mater. Sci. Eng. B **61–62**, 576–578 (1999)
134. G.S. Chung, Fabrication and characterization of a polycrystalline 3C-SiC piezoresistive micropressure sensor. J. Korean Phys. Soc. **56**(6), 1759–1762 (2010)
135. G. Wieczorek, B. Schellin, E. Obermeier, G. Fagnani, L. Drera, SiC based pressure sensor for high-temperature environments, in *Proceedings of the IEEE Sensors Conference* (2007), pp. 748–751
136. R.S. Okojie, A.A. Ned, A.D. Kurtz, Operation of α(6H)-SiC pressure sensor at 500°C. Sens. Actuators A-Phys. **66**, 200–204 (1998)
137. R.S. Okojie, D. Lukco, C. Blaha, V. Nguyen, E. Savrun, Zero offset drift suppression in SiC pressure sensors at 600°C, in *Proceedings of the IEEE Sensors Conference* (2010), pp. 2269–2274
138. R.S. Okojie, D. Lukco, V. Nguyen, E. Savrun, 4H-SiC piezoresistive pressure sensors at 800°C with observed sensitivity recovery. IEEE Electron Device Lett. **36**(2), 174–176 (2015)
139. G.S. Chung, R. Maboudian, Bonding characteristics of 3C-SiC wafers with hydrofluoric acid for high-temperature MEMS applications. Sens. Actuators A-Phys. **119**, 599–604 (2005)
140. F. Laermer, A. Urban, Challenges, developments and applications of silicon deep reactive ion etching. Microelectron. Eng. **67–68**, 349–355 (2003)
141. A.R. Atwell, R.S. Okojie, K.T. Kornegay, S.L. Roberson, A. Beliveau, Simulation, fabrication and testing of bulk micromachined 6H-SiC high-g piezoresistive accelerometers. Sens. Actuators A-Phys. **104**, 11–18 (2003)
142. M.A. Fraga, H. Furlan, S.M. Wakavaiachi, M. Massi, Fabrication and characterization of piezoresistive strain sensors for high temperature applications, in *Proceedings of the IEEE International Conference on Industrial Technology (ICIT)* (2010), pp. 513–516
143. M. Kumar, H. Bhaskaran, Ultrasensitive room-temperature piezoresistive transduction in graphene-based nanoelectromechanical systems. Nano Lett. **15**(4), 2562–2567 (2015)

Chapter 2
Theory of the Piezoresistive Effect in p-Type 3C-SiC

This chapter qualitatively explains the piezoresistive effect in p-type 3C-SiC based on the hole transfer mechanism and the conduction effective mass change due to the deformation of energy band under strain. To explain this phenomenon, the ideas of energy band structure and band deformation of 3C-SiC are discussed. Furthermore, the description of piezoresistive coefficients are also presented in the rest of this chapter.

2.1 Energy Band of 3C-SiC

2.1.1 Crystallographic and Energy Band Structure of 3C-SiC

The basic unit of 3C-SiC is based on the covalent bond between Si and C atoms, forming tetrahedrons. As such each Si atom bonds with other four C atoms to form SiC_4, while each C atom bonds with other four Si atoms to form C_4Si [1]. Different staking orders of these tetrahedrons result in different poly types of SiC, where the stacking sequence of these poly types can be presented using the 'ABC' notation, in which A, B and C represent the three sites available in one sub-lattice. For instance, the stacking sequence of a unit of 3C-SiC is ABC, while those of 4H-SiC and 6H-SiC are ABCB, and ABCACB, respectively [2]. Cubic SiC has been named as 3C-SiC, since each unit of its stacking order consists of three Si-C bi-layers, while the letter 'C' indicates that the symmetry of this poly type is cubic. The same rule has also been applied for other poly types such as 4 H and 6H-SiC in which numbers 4 and 6 indicate the number of Si-C bi-layers within one staking unit, and the letter 'H' stands for their hexagonal symmetry. Figure 2.1 shows a unit of SiC stacking tetrahedron and the stacking sequence (ABC) of 3C-SiC.

H.-P. Phan, *Piezoresistive Effect of p-Type Single Crystalline 3C-SiC*,
Springer Theses, DOI 10.1007/978-3-319-55544-7_2

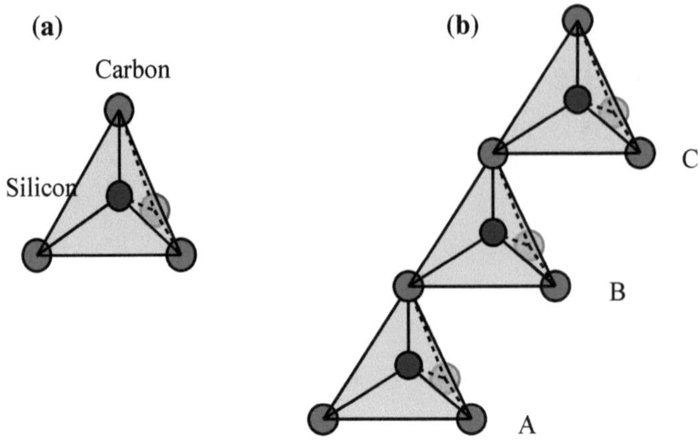

Fig. 2.1 **a** a unit of SiC_4 tetrahedron; **b** Stacking order of 3C-SiC

Figure 2.2a show the face centered crystal of 3C-SiC with its lattice constant of 4.36 Å [3]. Since semiconductor devices are built on or near the surface of semiconductor, the information of surface orientation are important. To define the orientation of a semiconductor plane, Miller indices are widely used [4]. The Miller indices of the most common planes in 3C-SiC are shown in Fig. 2.2b–d, including (100), (110), and (111) planes.

The band structure of a semiconductor is represented using the $E - k$ (Energy-momentum) diagram, where k is the momentum vector in the reciprocal space. The reciprocal lattice can be obtained from the real crystal lattice using a Fourier transformation and the band structure of a crystalline can be obtained by solving the Schrödinger equation of an one-electron problem [5]:

$$\left[-\frac{\hbar^2}{2m^*}\nabla^2 + V(r)\right]\Psi(r,k) = E(k)\Psi(r,k) \qquad (2.1)$$

where $\hbar = h/2\pi$ (i.e. $h = 6.626 \times 10^{-34}$ m^{2}kg/s is Plank's constant), m^* is the effective mass of electron, $V(r)$ is a potential energy in the direct lattice space, and $\Psi(r,k)$ is the wave function. Since the potential energy $V(r)$ is a periodic function, the energy $E(k)$ can be completely characterized in a single primitive cell in the reciprocal space. This primitive cell is defined as the first Brillouin zone (also known as the Brillouin zone), which can be constructed by drawing perpendicular bisector planes in the reciprocal lattice from the chosen center to the nearest equivalent reciprocal lattice sites. A detailed explanation of the reciprocal transformation and the construction of the Brillouin zone can be found in the book of S.M. Sze and Kwok K. Ng [4].

Figure 2.3 shows the first BZ of the face centered semiconductors, including Si and 3C-SiC. In this figure, the Γ point is the center of the BZ, the X point is the center of the square surfaces, and the L point is the center of the hexagonal surfaces,

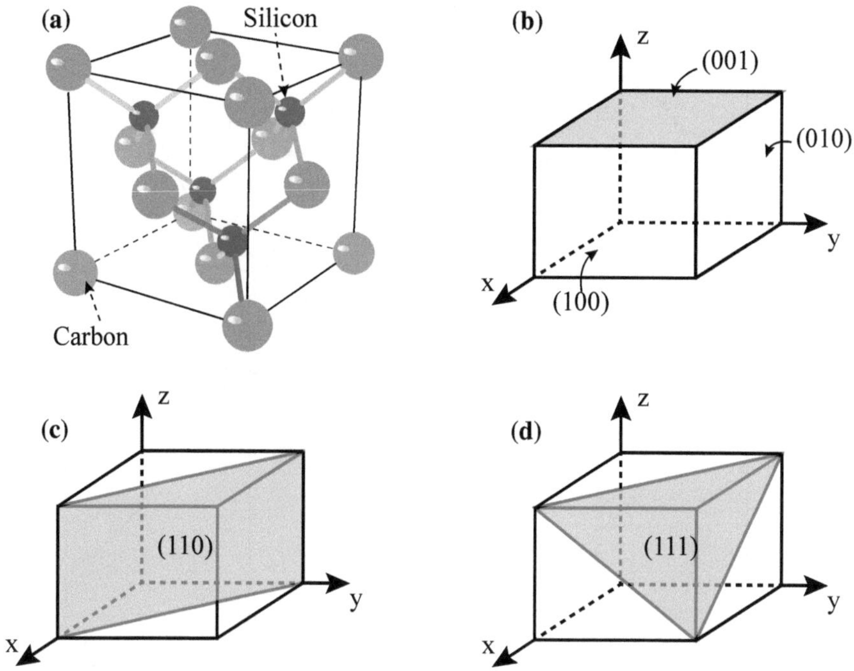

Fig. 2.2 **a** Crystal structure of 3C-SiC; **b**, **c**, **d** Common crystal planes of 3C-SiC

Fig. 2.3 The first Brillouin Zone (BZ) of face centered crystal

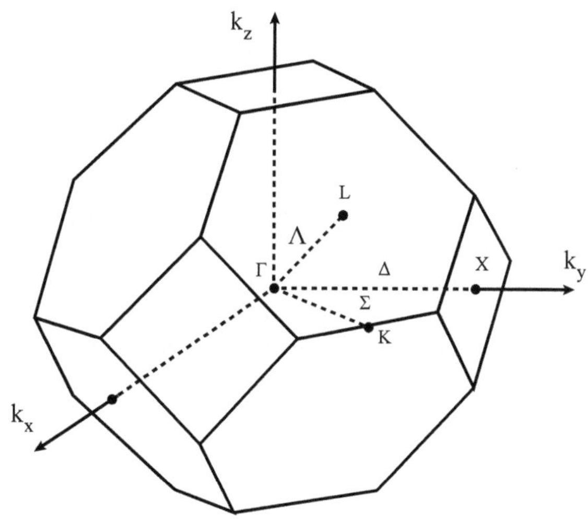

respectively. Additionally, Δ is the vector connecting the Γ point and the X point, while Λ is the vector connecting the Γ point and the L point.

The energy band of 3C-SiC was described in the paper of Hemstreet et al., calculated based on a non-local version of the empirical-pseudopotential method

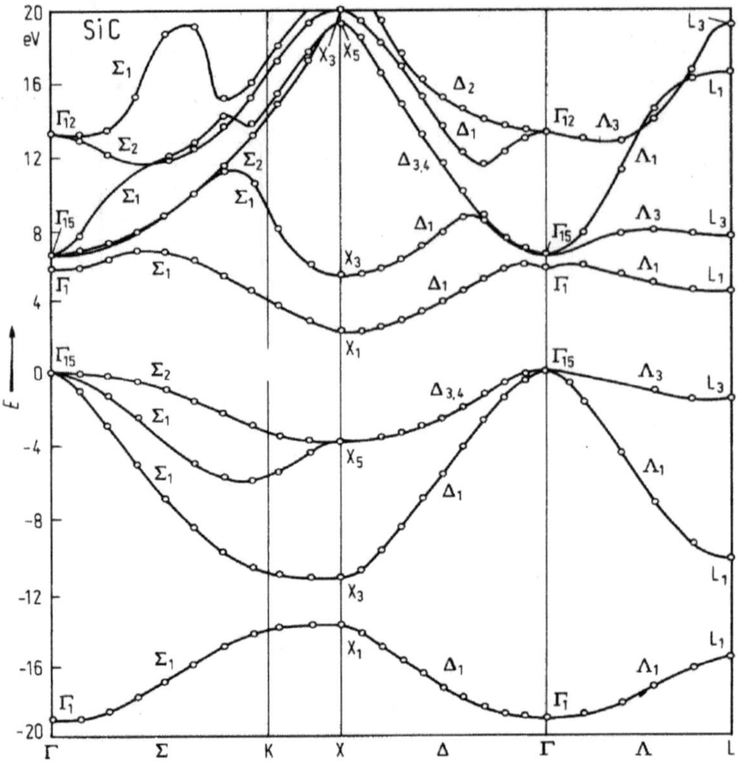

Fig. 2.4 The energy band structure of 3C-SiC in k-space. Reprint with permission from [6]

[6]. Accordingly, 3C-SiC is an indirect semiconductor, having its lowest conduction bands located at the X point, and its highest valance bands located at Γ point, as shown in Fig. 2.4. The indirect energy gap between X point and Γ point was calculated to be 2.33 eV, which is in consistent agreement with experimental results.

Since carriers (electrons in the conduction bands and holes in the valence bands) fill the lower energy bands first, the conductivity of 3C-SiC (σ_c) depends on the carrier concentration and mobility of the electrons in the bottom conduction band and the holes in the top valence band, which is defined in the following equation:

$$\sigma_c = q(\mu_e N_e + \mu_h N_h) \tag{2.2}$$

where q is the unit charge (i.e. 1.6021765×10^{19} C), μ_e and μ_h are the mobility of electron and hole, and N_e and N_h are the concentration of electrons and holes, respectively. In case of p-type 3C-SiC, holes are the major carriers, therefore its electrical conductivity is dominated by the hole mobility and concentration.

$$\sigma_c = q\mu_h N_h \tag{2.3}$$

where, the hole concentration (N_h) follows Boltzmann distribution:

$$N_h = N_V F_{1/2}\left(-\frac{E_F - E_V}{k_B T}\right) \tag{2.4}$$

where, N_V is the effective density of state, $F_{1/2}$ is the Fermi-Dirac integral, $E_F - E_V$ is the difference energy between the Fermi level and top valence bands, k_B is the Boltzmann's constant (i.e. 1.38×10^{-23} m^2.kg.s^{-2}.K^{-1}), and T is the absolute temperature, respectively. Furthermore, the mobility of hole is defined as:

$$\mu_h = \frac{q}{m_h^*}\tau_h \tag{2.5}$$

here m_h^* and τ_h are the hole effective mass and mean free path (i.e. the average time between two successive collisions of a particle to the lattice) of hole. Additionally, the effective mass can be expressed as a function of a wave-vector space:

$$m_h^* = \frac{h}{4\pi^2}\left(\frac{d^2E}{dk^2}\right)^{-1} \tag{2.6}$$

Equation 2.6 indicates that, in the $E - k$ diagram, a large energy curvature (parabolic approximation) results in a smaller hole effective mass. Therefore, in p-type 3C-SiC, the top two valance bands were named based on this properties, in which, the light hole has a smaller effective mass than heavy hole due to its larger energy curvature. Furthermore, substitute Eq. 2.5 into Eq. 2.3, the conductivity is:

$$\sigma_c = \frac{q^2 N_h \tau_h}{m_h^*} \tag{2.7}$$

Evidently, from Eqs. 2.4, 2.6, and 2.7, any change in band energy will result in a change of hole concentration (N_h) as well as hole effective mass (m_h^*), consequently changing the conductivity of p-type semiconductors. The change of these parameters under strain is presented in the following sections.

2.1.2 *Principle of the Piezoresistive Effect in p-Type 3C-SiC*

The piezoresistive effect in p-type semiconductor has been extensively investigated since the pioneering work of Smith in 1954 [7]. Its physical origin is considered to be the hole transfer and conduction mass shift mechanisms following the deformation of the top valence bands under strain. To understand the deformation of the top valence band, a simple model of one dimensional (1-D) lattice is introduced. Figure 2.5 shows the 1-D lattice model in which a non-interacting electron moves in a periodic 1-D array of potential wells created by positive ions arranged in a periodic order.

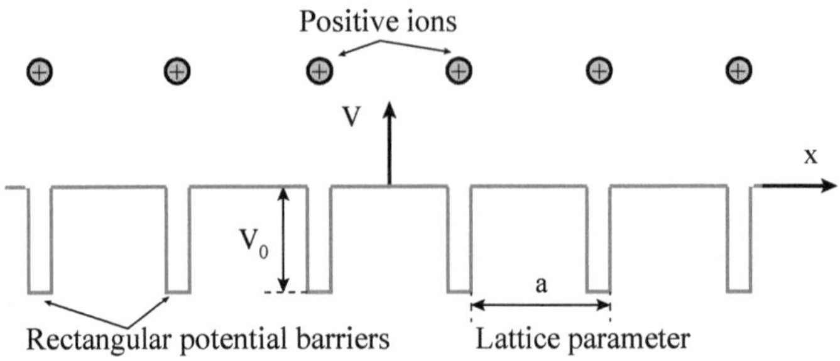

Fig. 2.5 Schematic sketch of 1-D lattice with approximated rectangular potentials

According to the approximation of Kronig and Penney, the potentials caused by the interaction between ions can be simplified as rectangular potential barriers spaced by the lattice parameter [8]. The solution of the Schrödinger's equation in this case is as follows [9]:

$$\cos(ka) = \cos(\alpha a) + \frac{m}{\hbar^2 \alpha} \sin(\alpha a) \tag{2.8}$$

where a is the lattice parameter, m is the free electron mass, and α is given by: $\alpha^2 = 2mE/\hbar^2$. Thus, at each wave vector k, the allowed energies (E) of the electron in the 1-D lattice are the solutions of Eq. 2.8.

Based on these solutions of E, the allowed energy bands of a non interacting electron in the 1-D lattice are plotted in Fig. 2.6 (the black solid curves) [9]. Additionally, from Eq. 2.8, by applying a mechanical strain to the 1-D lattices, the lattice size will be modified from a_0 to $(a_0 + \varepsilon a_0)$, resulting in a different solution of energy E at

Fig. 2.6 Energy band modification under mechanical strains in the 1-D lattice model [9]

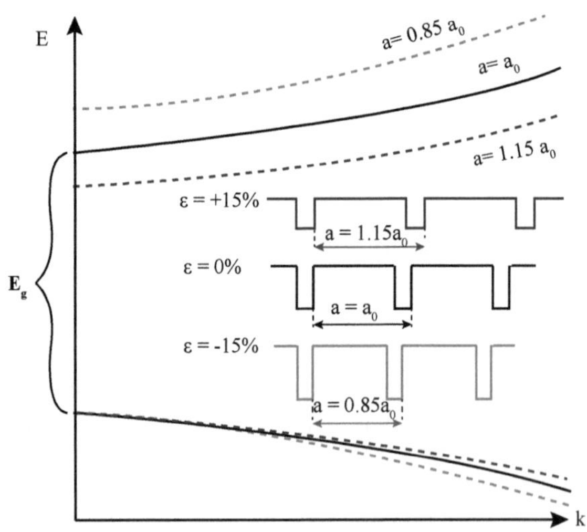

the wave vector k. As such, the simulation results reported by Rowe indicated that, applying strain of $\pm 15\%$ will cause the energy level (E) to shift up and down, leading to a change in the band gap between the allowed energy bands, as shown in the red and blue dashed lines of Fig. 2.6. These results indicate that the energy band structure of 1-D lattice can be modified by applying external mechanical strains.

For 3-D lattice crystals, the model is more complicated, and more complex calculation/simulation are required to estimate the stress dependence of the energy band structure. However, the principle is the same as the simple case of the 1-D lattice. Utilizing the theory of energy band modification under strain, the piezoresistive effect in semiconductors have been theoretically explained. Among various cubic crystalline semiconductors, the theory of the piezoresistive effect in Si has been extensively studied, and the theory of the strain effect on electrical conductivity in other materials has been developed by extending the model of Si. The next section will discuss the theory of the piezoresistance of p-type Si, based on that a qualitatively explanation of that phenomenon in p-type 3C-SiC is presented.

In the energy band structure of Si, two bands locate at the top valence, which are heavy hole and light hole bands. The spin-orbit split-off band locates far from the former bands (0.044 eV), therefore it can be negligible in the consideration of the electrical conduction in p-type Si. The band energies of heavy hole and light hole in Si are given in the following equation, using a parabolic approximation in which the electron effective mass is considered to be constant at the region near to the band extrema in the $E - k$ diagram [10]:

$$E(k) = E_{V0} + \frac{\hbar^2}{2m_0}\left\{Ak^2 \pm \sqrt{B^2k^4 + C^2(k_x^2k_y^2 + k_y^2k_z^2 + k_z^2k_x^2)}\right\} \tag{2.9}$$

where the + sign corresponds to the light hole, while the − sign corresponds to the heavy hole, E_{V0} is the top energy level in valence bands, and A, B, and C are inverse mass parameters, respectively. Under a mechanical strain, the lattice constants of Si and the symmetry of Si crystal will change. As such, a hydrostatic-compressive pressure will reduce its lattice size, while a tensile stress in [110] orientation will reshape Si crystal from a cubic to an orthorhombic structure. Since the lattice size and the symmetry of Si are changed under a mechanical strain/stress, its energy band is expected to be deformed, which also occurs in the 1-D lattice as discussed above. The deformation of the heavy and light holes was driven by Bir and Pikus [11], using the following equation:

$$E(\varepsilon, k) = E(0, k) \pm \frac{E_{\varepsilon k}}{2\sqrt{E_k}} \pm \frac{1}{\Lambda}E_{\varepsilon k} \tag{2.10}$$

where $E(0, k)$ is the strain-free energy of heavy hole and light hole presented in Eq. 2.9, while $E_{\varepsilon k}$ is an energy function of applied strain (ε) and wave vector (k), which is related to the strain tensor through a set of potential deformations (b and d) [12]:

$$E_{\varepsilon k} = Bb[3(k_x^2\varepsilon_{xx} + k_y^2\varepsilon_{yy} + k_z^2\varepsilon_{zz}) - k^2\varepsilon] + 3Dd(k_xk_y\varepsilon_{xy} + k_yk_z\varepsilon_{yz} + k_zk_x\varepsilon_{zx}) \tag{2.11}$$

here ε_{ij} are the six elements of the strain tensor in a 3-D coordinate. Consequently, the hole energy shift and the hole conduction mass shift under strain can be obtained as follows. The energy shift in heavy hole and light hole is:

$$\Delta E_{\varepsilon,k} = \pm\frac{E_{\varepsilon k}}{2\sqrt{E_k}} \tag{2.12}$$

and the conduction mass shift in these band is:

$$\Delta\frac{1}{m} = \pm\frac{1}{\Lambda}E_{\varepsilon k} \tag{2.13}$$

where the + sign corresponds to the heavy hole, and − sign corresponds to the light hole.

Figure 2.7 shows a schematic sketch of the splitting in heavy hole and light hole and the deformation of their energy surfaces under strain. Accordingly, given that a uniaxial strain is applied in [111] orientation of p-type Si, the heavy holes will shift up while the light hole will shift down. The energy surfaces of these bands are also warped due to the degeneracy. As such, the energy surfaces are elongated and become oblate ellipsoids, having axial symmetry along the direction of the applied strain [13].

Next, the electrical conductivity (or resistivity) of p-type Si under strain is considered. Neglecting the influence of spin-orbit split-off band, the conductivity of p-type Si is approximated as:

$$\sigma_{Si} = q^2\tau_h(p_{hh}/m_{hh} + p_{lh}/m_{lh}) \tag{2.14}$$

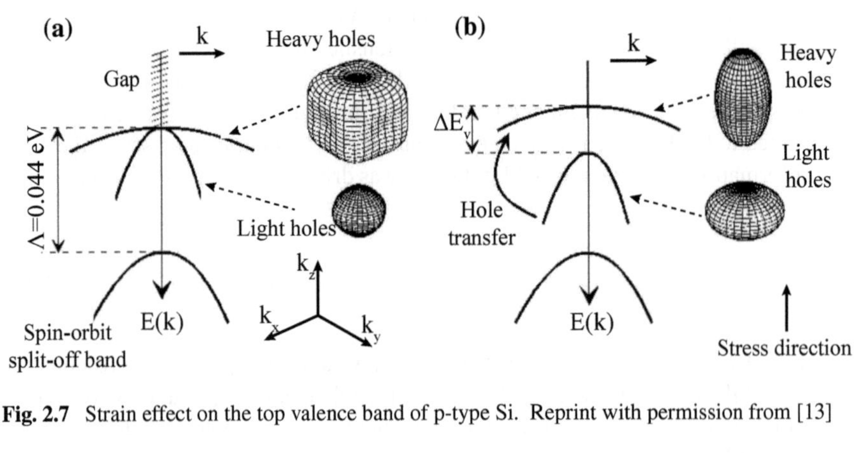

Fig. 2.7 Strain effect on the top valence band of p-type Si. Reprint with permission from [13]

where p and m are hole concentration and effective mass, while the subscription hh and lh indicate heavy hole and light hole.

Note that, since the concentration of hole follows the Boltzmann distribution, as presented in Eq. 2.4, the splitting and warping of heavy hole and light hole cause a change in the hole concentration in these two bands, following the rule that hole will move from higher energy levels to lower energy levels. This change in hole concentration can be quantified as:

$$\Delta P_{\mathrm{i}} = -\frac{N_{\mathrm{V,i}}\Delta(E_{\mathrm{V,i}} - E_{\mathrm{F}})}{k_{\mathrm{B}}T} F_{-1/2}(\eta_{\mathrm{F}}) \tag{2.15}$$

where $\eta_{\mathrm{F}} = -(E_{\mathrm{F}} - E_{\mathrm{V,i}})/(k_{\mathrm{B}}T)$ is the reduced Fermi energy [4], and $\Delta(E_{\mathrm{V,i}} - E_{\mathrm{F}})$ is the energy shift with respect to the Fermi level (E_{F}). Additionally, assume that the total number of holes remains constant, and hole transfer only occurs at the top valence bands (e.g. between heavy hole and light hole), hence:

$$\Delta P_{\mathrm{hh}} + \Delta P_{\mathrm{lh}} = 0 \tag{2.16}$$

Therefore, from Eqs. 2.15 and 2.16, the shift of the Fermi level is [12]:

$$\Delta E_{\mathrm{F}} = \frac{\sum_{i=1}^{2} \Delta E_{\mathrm{V,i}} N_{\mathrm{V,i}}}{\sum_{i=1}^{2} N_{\mathrm{V,i}}} F_{-1/2}(\eta_{\mathrm{F}}) \tag{2.17}$$

Consequently, the change in the concentration of heavy hole and light hole due to the shift of energy bands can be driven using the following equation:

$$\Delta P_{\mathrm{i}} = \frac{\Delta E_{\mathrm{V,i}} \sum_{i=1}^{2} N_{\mathrm{V,i}} - \sum_{i=1}^{2}(\Delta E_{\mathrm{V,i}} N_{\mathrm{V,i}})}{k_{\mathrm{B}}T \sum_{i=1}^{2} N_{\mathrm{V,i}}} F_{-1/2}(\eta_{\mathrm{F}}) \tag{2.18}$$

Based on Eqs. 2.13 and 2.18, the change of the conductivity in p-type Si is given by [14]:

$$\Delta\sigma_{\mathrm{Si},\varepsilon} = q^2\tau[\Delta p_{\mathrm{i}}\frac{1}{m_{\mathrm{i}}} + p_{\mathrm{i}}\Delta(\frac{1}{m_{\mathrm{i}}})] \tag{2.19}$$

Equation 2.19 indicates that both hole transfer (ΔP_{i}) and conduction mass shift ($\Delta(\frac{1}{m_{\mathrm{i}}})$) play an important role in the change of the electrical conductivity in p-type Si. Furthermore, experimental results and theoretical calculations reported in the literature also showed that the relative conductivity change of p-type Si per unit strain was approximately 100 which is two orders of magnitude larger than the change due to geometry. This indicated that the deformation of energy band structure in Si, resulting in the re-population of charge carrier and charge conduction mass, is the physical origin of the piezoresistive effect of p-type Si. Extending this established theory in Si, the piezoresistive effect of p-type 3C-SiC can be qualitatively presented as follows.

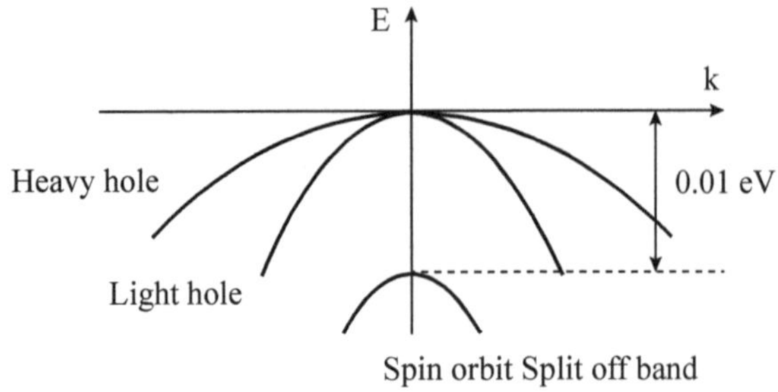

Fig. 2.8 Schematic sketch of the top valence band in p-type 3C-SiC

Because the top valence band of 3C-SiC also consists of three bands, similar to the case of Si, the piezoresistive effect in p-type 3C-SiC can be qualitatively explained using the model of Si. The difference between 3C-SiC and Si is that in 3C-SiC, the distance between the spin-orbit split-off band and the heavy/light holes is only 0.01 eV as shown in Fig. 2.8, the influence of this band should be taken into account [15]. This assumption was also made for the case of p-type diamond crystal, where the distance between spin-orbit split off band to the other two bands is also relatively small [16, 17].The conductivity of p-type 3C-SiC is given by [16]:

$$\sigma_{\mathrm{SiC}} = q^2\tau(P_{\mathrm{SiC,hh}}/m_{\mathrm{SiC,hh}} + P_{\mathrm{SiC,lh}}/m_{\mathrm{SiC,lh}} + P_{\mathrm{SiC,ss}}/m_{\mathrm{SiC,ss}}) \tag{2.20}$$

here the subscriptions hh, lh, and ss stand for heavy hole, light hole and spin-orbit split-off bands, respectively. Since the distribution of holes at the top valence bands and hole conduction mass depend on their energy levels as well as the interaction between the heavy hole, light hole, and and the spin-orbit split-off bands, any change in band energy structure induced by a mechanical strain could cause a change in the conductivity of p-type 3C-SiC (σ_{SiC}).

Numerous studies regarding the strain effect on energy band structure of 3C-SiC have been reported in the literature. Lu et al. used the linear muffin-tin orbitals (LMTO) method and the atomic-sphere approximation to investigate the influence of mechanical strain on the band structure of 3C-SiC [18]. The theoretical results showed that under a biaxial strain in (100) plane, energy band splitting could occur at L, X and Γ points. From the simulation results calculated based on the LMTO method, the energy splitting due to a biaxial strain at the Γ point (Δ_Γ) is approximated by the following equation:

$$\Delta E_\Gamma = b|\varepsilon_{\mathrm{xx}} - \varepsilon_{\mathrm{zz}}| + c(\varepsilon_{\mathrm{xx}} - \varepsilon_{\mathrm{zz}})^2 \tag{2.21}$$

where b and c are deformation potentials. In addition, the simulation results also showed that under a large bi-axial strain of above 8%, the direct gap at Γ point ($\Gamma_{15\mathrm{v}} - \Gamma_{1\mathrm{v}}$) will rapidly decrease in comparison to the indirect gap ($\Gamma_{15\mathrm{v}} - X_{1\mathrm{c}}$), turning 3C-SiC from an indirect material to a direct semiconductor. The authors, therefore, suggested that this interesting phenomenon could be of interest for the development of strain sensors and optical sensors. Rahimi et al. reported the bi-axial tensile stress on Al/p-type 3C-SiC/Al back to back Schottky diode [19]. Accordingly, applying a bi-axial stress in 3C-SiC will cause heavy hole and light hole in the valance band to split, where the heavy hole will be located higher in respect to the light hole. The split of these bands results in the change of band gap in 3C-SiC, thus changing the barrier of the Schottky diode. The change of this barrier can modify the current density of Al/SiC heterojunction.

The above-mentioned theoretical studies indicated that applying a mechanical strain could deform and shift the top valence bands in p-type 3C-SiC [18, 19]. As a result, this band deformation and shifting will change the hole concentration following the Boltzmann distribution (i.e. $P_{\mathrm{SiC,i}}$ becomes $P_{\mathrm{SiC,i}} + \Delta P_{\mathrm{SiC,i}}$), which is much the same as the case of Si described in the previous section. Additionally, the split of the top valance bands will result in the change in the interaction between heavy hole, light hole, and the spin-orbit split-off band, which causes the effective mass in these bands to be changed (i.e. $\frac{1}{m_{\mathrm{SiC,i}}}$ becomes $\frac{1}{m_{\mathrm{SiC,i}}} + \Delta\frac{1}{m_{\mathrm{SiC,i}}}$). Therefore, applying a strain is expected to significantly change the electrical conductivity of p-type 3C-SiC:

$$\frac{\Delta\sigma_{\mathrm{SiC},\varepsilon}}{\sigma_{\mathrm{SiC}}} = \frac{\sum_{i=1}^{3}\left(\Delta p_{\mathrm{SiC,i}}\frac{1}{m_{\mathrm{SiC,i}}} + p_{\mathrm{SiC,i}}\Delta\left(\frac{1}{m_{\mathrm{SiC,i}}}\right)\right)}{\sum_{i=1}^{3} p_{\mathrm{SiC,i}}\frac{1}{m_{\mathrm{SiC,i}}}} \tag{2.22}$$

where the subscription $i = 1, 2, 3$, stands for the heavy hole, light hole, and spin-orbit split-off bands, respectively. Due to the complexity of the valance band (the warping of heavy hole and light hole, as well as the strong interaction between these bands with the spin-orbit split off band), to date these has been no quantitative estimation/calculation of the piezoresistive effect in p-type 3C-SiC. Therefore, more theoretical studies need to be carried out in order to theoretically understand the physical origin of the strain effect in the electrical conduction of p-type 3C-SiC. Additionally, in this thesis, the experimental approach was utilized to address the magnitude, the orientation dependence, and the temperature dependence of the piezoresistance of p-type 3C-SiC. The experimental data obtained in this work will be significant for the perspective theoretical work on the physical phenomenon of strain effect in p-type 3C-SiC.

2.2 The Piezoresistive Coefficients of 3C-SiC

2.2.1 Definition of the Piezoresistive Coefficients

As mentioned in Chap. 1, the relative resistivity change of a 3C-SiC resistance caused by a mechanical strain can be expressed as a function of the gauge factor (GF) of a semiconductor and the applied strain:

$$\frac{\Delta\rho}{\rho} = GF \times \varepsilon \tag{2.23}$$

In addition, this change of resistivity ($\Delta\rho/\rho$) is connected to the applied stress (σ) through a parameter named piezoresistive coefficient (π):

$$\Delta\rho/\rho = \pi\sigma \tag{2.24}$$

Furthermore, when a uniaxial stress is applied, the relationship between the stress and strain follows Hooke's law: $\sigma = E\varepsilon$, where E is the Young's modulus of SiC. Thus, the relationship between the gauge factor and the piezoresistive coefficient is $GF = E\pi$ [20].

In a general case, where stresses are applied in arbitrary crystallography orientations as shown in Fig. 2.9, the change of resistivity $\Delta\rho_{\mathrm{ij}}$ is a second rank tensor, which is connected to stress tensors (σ_{kl}) by a forth-rank piezoresistive coefficient tensor (π_{ijkl}) [21]:

$$\frac{\Delta\rho_{\mathrm{ij}}}{\rho} = \sum_{\mathrm{k,l}} \pi_{\mathrm{ijkl}}\sigma_{\mathrm{kl}} \tag{2.25}$$

where, i and j denote the directions of the applied current and voltage, while k and l indicate the orientations of the applied stress tensor, respectively. The forth-rank piezoresistive coefficient tensor can be collapsed to a second rank tensor (e.g. $\pi_{1111} \rightarrow \pi_{11}$, $\pi_{1122} \rightarrow \pi_{12}$, and $\pi_{2323} \rightarrow \pi_{44}$), using the transformation scheme shown in Table 2.1 [20]. Equation 2.25 and Table 2.1 also show that in a three dimensional coordinate, the number of stress components is 6, while the number of piezoresistive coefficients is $6 \times 6 = 36$. However, since 3C-SiC has a symmetric crystalline structure, the number of basic piezoresistive coefficients reduces to 3 in the principal coordinate. The relationship and the applied stress tensors in single crystalline 3C-SiC can be represented in a matrix form as follows:

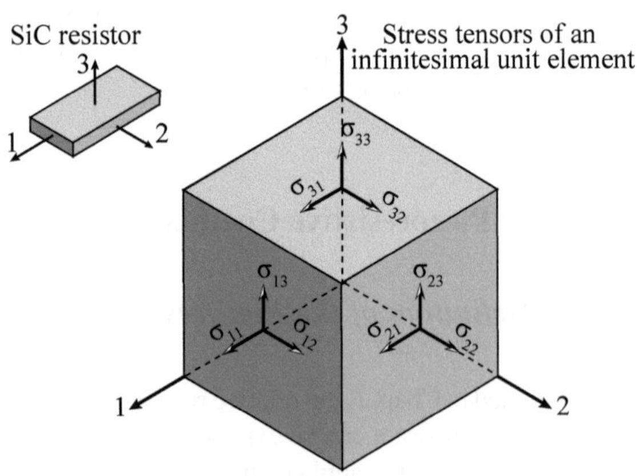

Fig. 2.9 The nine stress components σ_{kl} of an infinitesimal unit element of SiC

Table 2.1 Index transformation scheme

Tensor notation	11	22	33	23 and 32	13 and 31	12 and 21
Matrix notation	1	2	3	4	5	6

$$\begin{pmatrix} \frac{\delta\rho_{11}}{\rho_{11}} \\ \frac{\delta\rho_{22}}{\rho_{22}} \\ \frac{\delta\rho_{33}}{\rho_{33}} \\ \frac{\delta\rho_{23}}{\rho_{23}} \\ \frac{\delta\rho_{13}}{\rho_{13}} \\ \frac{\delta\rho_{12}}{\rho_{12}} \end{pmatrix} = \begin{pmatrix} \pi_{11} & \pi_{12} & \pi_{12} & 0 & 0 & 0 \\ \pi_{12} & \pi_{11} & \pi_{12} & 0 & 0 & 0 \\ \pi_{12} & \pi_{12} & \pi_{11} & 0 & 0 & 0 \\ 0 & 0 & 0 & \pi_{44} & 0 & 0 \\ 0 & 0 & 0 & 0 & \pi_{44} & 0 \\ 0 & 0 & 0 & 0 & 0 & \pi_{44} \end{pmatrix} \begin{pmatrix} \varepsilon_{11} \\ \varepsilon_{22} \\ \varepsilon_{33} \\ \varepsilon_{23} \\ \varepsilon_{13} \\ \varepsilon_{12} \end{pmatrix} \tag{2.26}$$

The fundamental piezoresistive coefficients π_{11}, π_{12}, and π_{44} are very important in defining the orientations dependence of piezoresistive effect, as well as to determine which orientation offers the largest sensitivity for mechanical sensing devices [22].

2.2.2 Piezoresistive Coefficients of Two-Terminal and Four-Terminal Resistors

There are two types of piezoresistors which are widely deployed in MEMS sensors [23]. The most common type is the two-terminal resistor where the directions of applied electrical field vector (E) and the current density vector (J) are the same, Fig. 2.10a. The relationship between the change of resistivity and applied stress can be described as follows:

$$\frac{\Delta\rho}{\rho} = \pi'_{11}\sigma_1 + \pi'_{12}\sigma_2 + \pi'_{13}\sigma_3 + \pi'_{14}\sigma_4 + \pi'_{15}\sigma_5 + \pi'_{16}\sigma_6 \tag{2.27}$$

where, σ_1, σ_2, and σ_3 are normal stresses, while σ_4, σ_5, and σ_6 is a shear stress which corresponds. The prime superscription denotes the piezoresistive coefficients in the

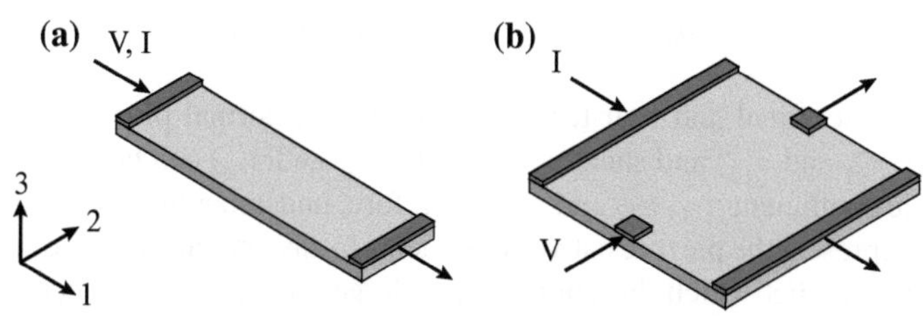

Fig. 2.10 Concept of SiC resistors. **a** Two-terminal resistor; **b** Four-terminal resistor

coordinate of the piezoresistor but not in the principal coordinate. For the case of piezoresistors fabricated using thin films, the piezoresistive coefficient components in the direction perpendicular to applied electric field and current (also known as the out of plane direction) are negligible, because the influence of the stress components in this direction is considered to be insignificant. Therefore, the relative resistance change of thin film resistors under stress is deduced as:

$$\frac{\Delta\rho}{\rho} = \pi'_{11}\sigma_1 + \pi'_{12}\sigma_2 + \pi'_{16}\sigma_6 \tag{2.28}$$

Given that the direction of applied current is the longitudinal orientation, while the direction perpendicular to current is the transverse orientation, Eq. 2.28 can be represented as:

$$\frac{\Delta\rho}{\rho} = \pi_l\sigma_l + \pi_t\sigma_t + \pi_s\sigma_s \tag{2.29}$$

where the subscriptions l, t, and s indicate longitudinal, transverse, and shear components.

The second type of piezoresistors is four-terminal devices [24, 25], Fig. 2.10b. The piezoresistive effect in these devices have attracted considerable attention, since it can be deployed to investigate the influence of stress on the offset voltage in Hall devices. The principle of the piezoresistive effect in two-terminal resistors is that, when a stress is applied to the device, a voltage will be generated at two electrodes which are are aligned in a direction perpendicular to the supplied current. Since this voltage is generated due to strain/stress, without requiring an external magnetic field like the case of Hall devices, the piezoresistive effect in two-terminal resistors is also called the pseudo-Hall effect. Additionally, the ratio between the generated voltage and applied current is defined as a pseudo resistance ρ_{pseudo}, which is connected to the applied stress tensor through the shear piezoresistive coefficients [24, 26]:

$$\frac{V}{I} = \rho_{\text{pseudo}} = \pi'_{61}\sigma_1 + \pi'_{62}\sigma_2 + \pi'_{63}\sigma_6 + \pi'_{64}\sigma_4 + \pi'_{65}\sigma_5 + \pi'_{66}\sigma_6 \tag{2.30}$$

Similar to the case of thin film two-terminal resistors, for thin film four-terminal resistors, the stress components related to the out of plane direction are negligible. Consequently, the generated pseudo resistance can be approximated as:

$$\rho_{\text{pseudo}} = \pi'_{61}\sigma_1 + \pi'_{62}\sigma_2 + \pi'_{66}\sigma_6 \tag{2.31}$$

For both two-terminal and four-terminal devices, the normal piezoresistive coefficients (e.g. π'_{11} and π'_{12}) and shear coefficients (π'_{61} and π'_{62}) can be estimated based on the basic coefficient π_{11}, π_{12} and π_{44}. Therefore, understanding these basic coefficients can represent the piezoresistive coefficients in any crystallographic orientation, and thus can predict which directions offer a large piezoresistive effect. The transformation of piezoresistive coefficient in arbitrary Cartesian coordinate is presented in the following subsection.

2.2.3 *Piezoresistive Coefficients in Arbitrary Cartesian Coordinate*

Given that X, Y, Z are the principal axis, and X', Y', Z' are the coordinate of a piezoresistor, the Cartesian coordinate of the piezoresistor can be obtained from the principal coordinate using the axes transformation illustrated in Fig. 2.11. The first rotation is by an angle ϕ about Z axis; the second rotation is about Y_1 axis an angle of θ; and the third rotation is about Z_2 axis an angle of φ, as illustrated in Fig. 2.11. Additionally, the coordinate x', y', z' in the coordinate of piezoresistor can be transformed from the coordinate x, y, z displayed in the principle Cartesian system, using the following equation:

$$\begin{pmatrix} x' \\ y' \\ z' \end{pmatrix} = \begin{pmatrix} l_1 & m_1 & n_1 \\ l_2 & m_2 & n_2 \\ l_3 & m_3 & n_3 \end{pmatrix} \begin{pmatrix} x \\ y \\ z \end{pmatrix} \tag{2.32}$$

Here l_i, m_j, and n_k (where, i, j, k $\in \{1, 2, 3\}$) are directional cosine, which are calculated from the rotation angles ϕ, θ, and φ, using the Euler rotation matrix:

$$\begin{pmatrix} l_1 & m_1 & n_1 \\ l_2 & m_2 & n_2 \\ l_3 & m_3 & n_3 \end{pmatrix} = \begin{pmatrix} \cos\phi\cos\theta\cos\varphi - \sin\phi\cos\varphi & \sin\phi\cos\theta\cos\varphi + \cos\phi\sin\varphi & -\sin\phi\cos\varphi \\ -\cos\phi\cos\theta\sin\varphi - \sin\phi\cos\varphi & -\sin\phi\cos\theta\sin\varphi + \cos\phi\cos\varphi & \sin\theta\sin\varphi \\ \cos\phi\sin\theta & \sin\phi\sin\theta & \cos\theta \end{pmatrix} \tag{2.33}$$

Using Euler's rotation theory, the piezoresistive coefficient tensors in the Cartesian system of a piezoresistor can be calculated from the principle piezoresistive coefficient tensor in the principal coordinate, using the following equation [27]:

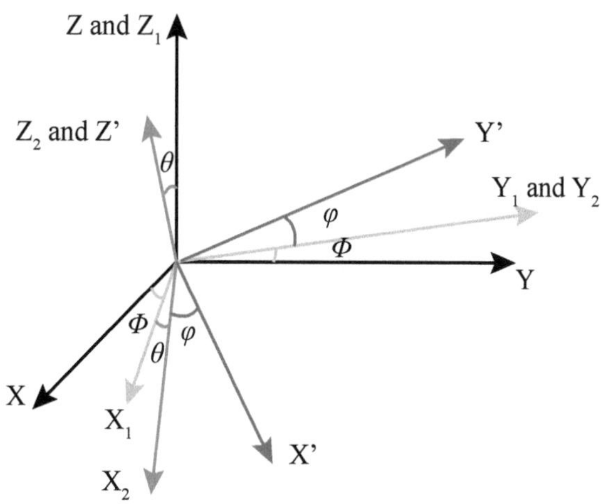

Fig. 2.11 Euler's rotation angles

Table 2.2 Piezoresistive coefficients in arbitrary coordinate

Coeff.	Expressions	Coeff.	Expressions
π'_{11}	$\pi_{11} - 2\pi_C(l_1^2m_1^2 + m_1^2n_1^2 + n_1^2l_1^2)$	π'_{61}	$\pi_C(l_1^3l_2 + m_1^3m_2 + n_1^3n_2)$
π'_{12}	$\pi_{12} + \pi_C(l_1^2l_2^2 + m_1^2m_2^2 + n_1^2n_2^2)$	π'_{62}	$\pi_C(l_1l_2^3 + m_1m_2^3 + n_1n_2^3)$
π'_{13}	$\pi_{12} + \pi_C(l_1^2l_3^2 + m_1^2m_3^2 + n_1^2n_3^2)$	π'_{63}	$\pi_C(l_1l_2l_3^2 + m_1m_2m_3^2 + n_1n_2n_3^2)$
π'_{14}	$2\pi_C(l_1^2l_2l_3 + m_1^2m_2m_3 + n_1^2n_2n_3)$	π'_{64}	$2\pi_C(l_1l_2^2l_3 + m_1m_2^2m_3 + n_1n_2^2n_3)$
π'_{15}	$\pi_C(l_1^3l_3 + m_1^3m_3 + n_1^3n_3)$	π'_{65}	$2\pi_C(l_1^2l_2l_3 + m_1^2m_2m_3 + n_1^2n_2n_3)$
π'_{16}	$2\pi_C(l_1^3l_2 + m_1^3m_2 + n_1^3n_2)$	π'_{66}	$\pi_{44} + 2\pi_C(l_1^2l_2^2 + m_1^2m_2^2 + n_1^2n_2^2)$

$\pi_C \equiv \pi_{11} - \pi_{12} - \pi_{44}$

$$(\pi_{ij}) = \alpha^{-1}\begin{pmatrix} \pi_{11} & \pi_{12} & \pi_{12} & 0 & 0 & 0 \\ \pi_{12} & \pi_{11} & \pi_{12} & 0 & 0 & 0 \\ \pi_{12} & \pi_{12} & \pi_{11} & 0 & 0 & 0 \\ 0 & 0 & 0 & \pi_{44} & 0 & 0 \\ 0 & 0 & 0 & 0 & \pi_{44} & 0 \\ 0 & 0 & 0 & 0 & 0 & \pi_{44} \end{pmatrix}\alpha \quad (2.34)$$

where:

$$\alpha = \begin{pmatrix} l_1^2 & m_1^2 & n_1^2 & 2m_1n_1 & 2n_1l_1 & 2l_1m_1 \\ l_2^2 & m_2^2 & n_2^2 & 2m_2n_2 & 2n_2l_2 & 2l_2m_2 \\ l_3^2 & m_3^2 & n_3^2 & 2m_3n_3 & 2n_3l_3 & 2l_3m_3 \\ l_2l_3 & m_2m_3 & n_2n_3 & m_2n_3 + m_3n_2 & n_2l_3 + n_3l_2 & m_2l_3 + l_3m_2 \\ l_3l_1 & m_3m_1 & b_3n_1 & m_3n_1 + m_1n_3 & n_3l_1 + n_1l_3 & m_3l_1 + l_1m_3 \\ l_1l_2 & m_1m_2 & b_1n_2 & m_1n_2 + m_2n_1 & n_1l_1 + n_2l_1 & m_1l_2 + l_2m_1 \end{pmatrix} \quad (2.35)$$

Based on Eq. 2.35, the transformation of the piezoresistive coefficients of two-terminal and four-terminal resistors in any crystallographic orientation are presented in Table 2.2 [27].

References

1. D. Olego, M. Cardona, P. Vogl, Pressure dependence of the optical phonons and transverse effective charge in 3 C-SiC. Phys. Rev. B **25**(6), 3878 (1982)
2. M. Wijesundara, R. Azevedo, *Silicon Carbide Microsystems for Harsh Environments* (Springer Science & Business Media, Berlin, 2011)
3. G.L. Harris, *Properties of Silicon Carbide*, vol. 13 (IET, London, 1995)
4. S.M. Sze, K.K. Ng, *Physics of Semiconductor Devices* (Wiley, New York, 2006)
5. J.M. Ziman, *Principle of the Theory of Solids* (Cambridge University Press, London, 1964)
6. L.A. Hemstreet Jr., C.Y. Fong, Electronic band structure and optical properties of 3C-SiC, BP, and BN. Phys. Rev. B **6**(4), 1464 (1972)
7. W.P. Eaton, J.H. Smith, Micromachined pressure sensors: review and recent developments. Smar. Mater. Struct. **6**, 530–539 (1997)
8. J.C. Wolfe, Summary of the Kronig–Penney electron. Am. J. Phys. **46**, 1012 (1978)

9. A.C.H. Rowe, Piezoresistance in silicon and its nanostructures. J. Mater. Res. **29**(6), 731–744 (2014)
10. D.L. Pulfrey, *Understanding Modern Transistors and Diodes* (Cambridge University Press, London, 2010)
11. G.L. Bir, G.E. Pikus, in *Symmetry and Strain-Induced Effects in Semiconductors*, vol. 624, ed. by D. Louvish (Wiley, New York, 1974)
12. T. Toriyama, S. Sugiyama, Analysis of piezoresistance in p-type silicon for mechanical sensors. J. Microelectromech. Syst. **11**(5), 598 (2002)
13. P. Kleimann, B. Semmache, M. Le Berre, D. Barbier, Stress-dependent hole effective masses and piezoresistive properties of p-type monocrystalline and polycrystalline silicon. Phys. Rev. B **57**(15), 8966–8971 (1998)
14. Y. Sun, S.E. Thompson, T. Nishida, *Strain Effect in Semiconductor: Theory and Device Applications*, 1st edn. (Springer, Berlin, 2009)
15. H.-P. Phan, D.V. Dao, K. Nakamura, S. Dimitrijev, N.-T. Nguyen, The piezoresistive effect of SiC for MEMS sensors at high temperatures: a review. J. Microelectromech. Syst. **24**(6), 1663–1677 (2015)
16. C. Kong, W. Wang, K. Liao, Y. Ma, S. Wang, L. Fang, The theoretical studies of piezoresistive effect in diamond films. Sci. China Ser. A Math. **45**(1), 107–114 (2002)
17. L. Fang, W.L. Wang, P.D. Ding, K.J. Liao, J. Wang, Study on the piezoresistive effect of crystalline and polycrystalline diamond under uniaxial strains. J. Appl. Phys. **86**, 5185 (1999)
18. W. Lu, K. Zhang, X. Xe, Strain effects on the band structures of β-SiC. Phys. Condens. Matter **5**, 883–890 (1993)
19. R. Rahimi, C.M. Miller, S. Raghavan, C.D. Stinespring, D. Korakakis, Electrical properties of strained nano-thin 3C-SiC/Si heterostructures. J. Phys. D Appl. Phys. **42**, 055108 (2009)
20. A.A. Barlian, W.T. Park, J.R. Mallon Jr., A.J. Rastergar, B.L. Pruit, Review: semiconductor the piezoresistive effect for microsystems. Proc. IEEE **97**(3), 513–552 (2009)
21. Y. Kanda, The piezoresistive effect of silicon. Sens. Actuators A **28**, 83–91 (1991)
22. Y. Kanda, A graphical representation of the piezoresistance coefficients in silicon. IEEE Trans. Electron Devices **29**(1), 64–70 (1982)
23. J.C. Doll, B.L. Pruitt, *Piezoresistor Design and Applications* (Springer, Berlin, 2013). ISBN 978-1-4614-8516-2
24. Y. Kanda, K. Yamamura, Four terminal gauge quasi circular and square diaphragm silicon pressure sensors. Sens. Actuators **19**, 247–257 (1989)
25. A. Mian, J.C. Suhling, R.C. Jaeger, The van der Pauw Stress Sensor. IEEE Sens. J. **6**(2), (2006)
26. D.V. Dao, T. Toriyama, J. Wells, S. Sugiyama, Silicon piezoresistive six-degree of freedom micro force-moment sensor. Sens. Mater. **15**, 113–135 (2002)
27. M.H. Bao, *Micro Mechanical Transducers: Pressure Sensors, Accelerometers and Gyroscopes*, vol. 8 (Elsevier, Amsterdam, 2000)

Chapter 3
3C-SiC Film Growth and Sample Preparation

This chapter presents the growth process of p-type single crystalline 3C-SiC, the optical characterization and the electrical properties of 3C-SiC films. The fabrication of SiC resistors used for investigating the piezoresistance in p-type 3C-SiC is also described.

3.1 Growth of Single Crystalline 3C-SiC

3.1.1 The LPCVD Process of p-Type 3C-SiC

Figure 3.1 represents the growth process of p-type 3C-SiC on a Si(100)wafer at Queensland Micro and Nanotechnology Centre using a hot-wall LPCVD reactor at 1000 °C [1, 2]. The alternating supply epitaxy approach was employed to deposit single crystalline SiC with silane (SiH_4) and propylene (C_3H_6) being used as the precursors. Compared to the conventional heteroepitaxial growth of 3C-SiC using LPCVD where silicon-containing and carbon-containing precursors are simultaneously supplied, the alternating supply growth significantly reduced the growth temperature (1000° compared to 1350° using simultaneous methods). This low growth temperature not only reduce the dopant redistribution in Si substrate, but also, reduces the wafer bow, which enables the growth of 6-inch 3C-SiC/Si wafers. Additionally, to form the p-type semiconductors, boron (B) and aluminum (Al) are usually used. Compared to B, Al has a shallower acceptor level and is much less influenced by hydrogen incorporation during chemical vapor deposition growth [2]. Therefore, in this study, Al was selected as the impurities for p-type 3C-SiC, and Trimethylaluminium [$(CH_3)_3Al$, TMAl] was used for the *in situ* doping process.

Prior to the epitaxial deposition, the Si surface was cleaned using the standard Radio Corporation America (RCA) cleaning procedures. Subsequently, the idle temperature of the hot-wall chamber was set at 600 °C, and after loading the Si wafer, the reactor was ramped up from 600 to 1000 °C. Subsequently, the LPCVD process

H.-P. Phan, *Piezoresistive Effect of p-Type Single Crystalline 3C-SiC*,
Springer Theses, DOI 10.1007/978-3-319-55544-7_3

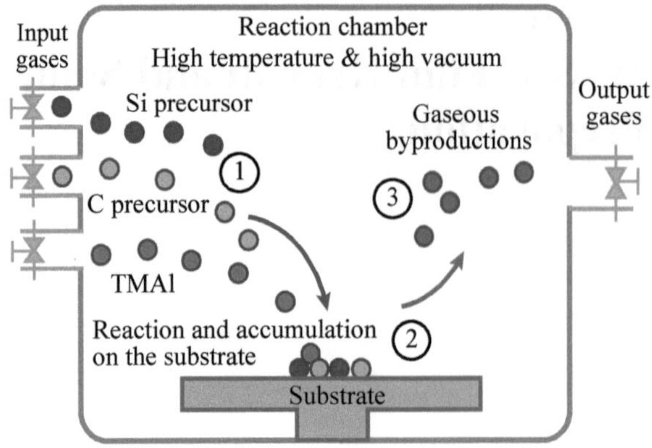

Fig. 3.1 Concept of a CVD growth process

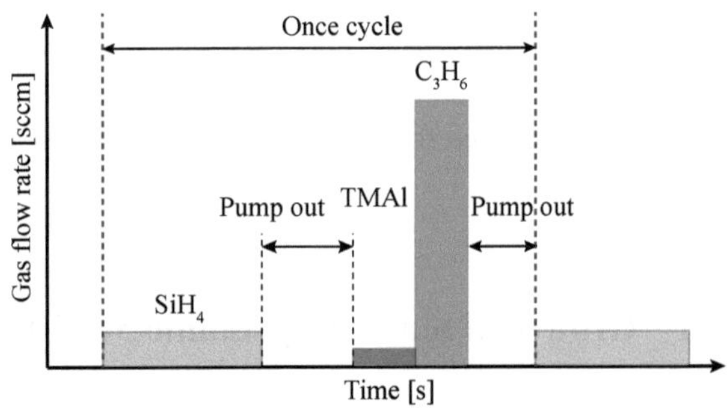

Fig. 3.2 Schematic diagrams of the gas supply procedure of the LPCVD process

was performed with the following five steps and these steps were repeated for several cycles, as illustrated in Fig. 3.2:

(i) Supply of SiH_4 for 10 s with a flow rate of 2 standard cubic centimeter per minute (sccm)
(ii) Pump out for 30 s
(iii) Supply TMAl (0.1 sccm) for 10–20 s (*in situ* doping)
(iv) Supply 1.5 sccm C_3H_6
(v) Pump out for 30 s.

The growth rate was controlled at 0.61 ± 0.02 nm/cycle and the thickness of the SiC layer can be modified by changing the number of growth cycles.

Figure 3.3 shows 3C-SiC films with different thicknesses. Evidently, SiC/Si wafers with different thicknesses of the 3C-SiC layer show different colors when being observed with naked eyes or optical microscopes. For instance, the color of a 80 nm 3C-SiC film is light steel blue, while those of 130 and 280 nm SiC films are goldenrod and green, respectively. A detailed description on the color chart of 3C-SiC/Si wafer was presented in [3].

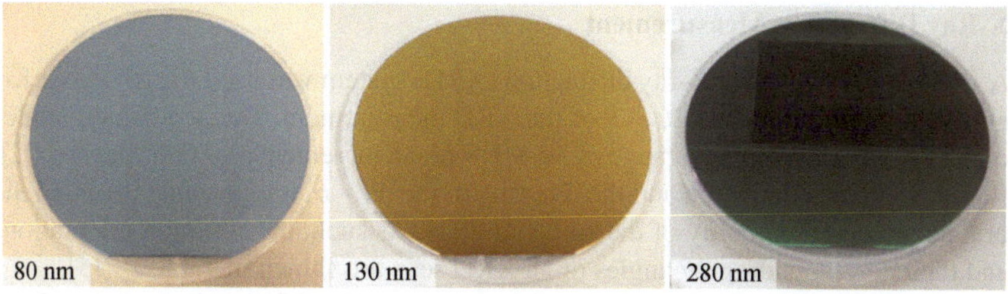

Fig. 3.3 3C-SiC on Si wafers with different thicknesses grown by the alternatively supplying CVD. The diameter of all wafers used in this work was 150 mm. The thicknesses of these SiC films were measured by NANOMETRICS Nanospec/AFT 210

3.1.2 *Optical Characterization of the As-Grown 3C-SiC Film*

The properties of the as-grown 3C-SiC films were investigated by Atomic Force Microscopy (AFM), X-Ray Diffractometry (XRD), Transmission Electron Microscopy (TEM), and selected area electron diffraction (SAED).

Surface Roughness Measurement

The surface topography of 3C-SiC film was characterized using AFM (model: MFP3D-BIO) where the operating mode was in air tapping and the radius of the curvature of the cantilever tip was 10 nm. Figure 3.4 shows the AFM image of a 5 μm × 5 μm area of SiC film having a thickness of 280 nm. The root mean square (RMS) of the SiC surface was found to be below 20 nm [4].

Fig. 3.4 The atomic force microscope (AFM) image of 3C-SiC

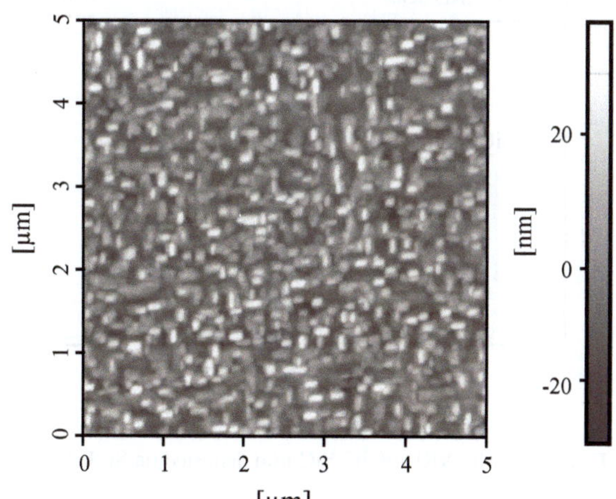

X-Ray Diffraction Measurement

A ™Bruker D8 advance X-ray diffraction (XRD) system was used to provide CuKα emission in the full-range $2\theta - \theta$ of the XRD measurement. The acquisition angles of the $2\theta - \theta$ scan were in a range of 30–90° with an increment of 0.005° per step [2]. Based on the peak position and the intensity of the $2\theta - \theta$, it is possible to obtain the orientation and the quality of the epitaxial films. As such, the common orientations and their corresponding 2θ angles of an XRD scan are shown in Table 3.1 [5].

Accordingly, from the $2\theta - \theta$ scan of the as-grown 3C-SiC film shown in Fig. 3.5a, diffraction peaks at 35.6° and 90° which correspond to 3C-SiC(200) and 3C-SiC(400) orientations, were observed. In addition, besides these two peaks, only a peak at 69.1° was detected, corresponding to Si(400) [6]. This result indicated that single crystalline (100) 3C-SiC was grown on Si substrate and the orientations of 3C-SiC film and the Si substrate were the same. The full width at half maximum (FWHM) of SiC (200) peak in the $2\theta - \theta$ scan is 0.26° [7].

To further examine the crystal quality, the full width at half maximum (FWHM) of rocking curve analysis of a $2\theta - \omega$ scan was employed, Fig. 3.5b. Accordingly, the FWHM of rocking curve scan of the 3C-SiC(200) peak was found to be 0.80° which was comparable to other 3C-SiC films grown by LPCVD [8].

Table 3.1 List of miller indices and corresponding diffraction angles of 3C-SiC [5]

2θ	35.6°	41.4°	60°	71.8°	75.5°	90°
Orientation [hkl]*	111	200	220	311	222	400

*[hkl] indicates the Miller indices

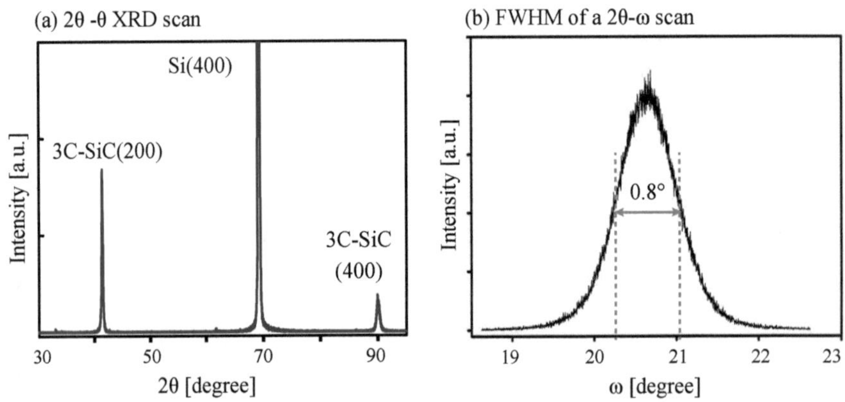

Fig. 3.5 **a** The XRD of 3C-SiC film deposted on Si; **b** The rocking curve scan of 3C-SiC

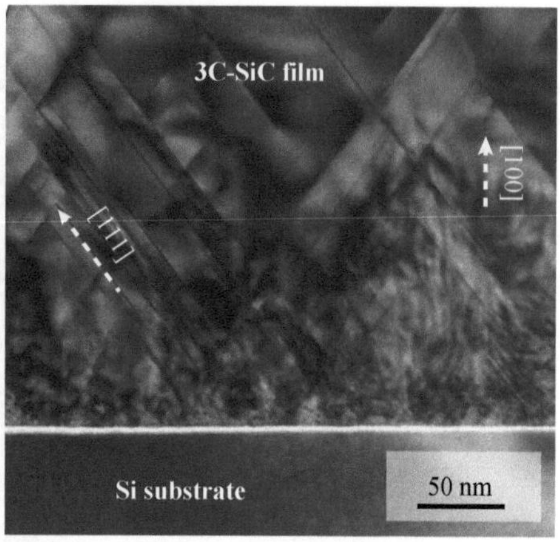

Fig. 3.6 The TEM image of the as-grown 3C-SiC

Transmission Electron Microscopy and Selected Area Electron Diffraction Measurements

The types of crystal defects and their density were characterized using TEM. Figure 3.6 shows a cross-sectional TEM image of a 280 nm 3C-SiC film along [110] direction of electron incidence. It can be seen that there are no boundaries in the single crystalline 3C-SiC, and staking faults propagating in [111] direction are the dominant crystal defects, which are densely contributed at the SiC/Si interface. Additionally, it is also evident from the TEM image that, the quality of the SiC film increased in the layers locating significantly well-above the SiC/Si interface [9].

Furthermore, the selected area electron diffraction (SAED) pattern observed in the TEM measurement can also be used to identify crystal structures and examine crystal defects. Typically, there are three types of SAED patterns, which are (i) spots arranged in a periodic order indicating a crystalline material, (ii) diffuse rings indicating amorphous material, and the small spots making up a ring implying poly crystalline material [10]. Figure 3.7 represents the SEAD from both the SiC film and the Si substrate observed along the [110] orientation. Consequently, the bright spots observed in the SAED image confirm that the grown 3C-SiC is single crystalline.

3.1.3 *Electrical Properties of the As-Grown 3C-SiC Films*

The semiconductor type of the films was measured using a hot probe technique. The measurement was performed by contacting the 3C-SiC films to a hot probe and a cold probe, where the hot probe was grounded to minimize any noise from the heater, as illustrated in Fig. 3.8. When these two probes are in contact with the films, the carriers of the films will flow from high to low temperatures, as a results of the electric field generated between the hot and cold probes.

Fig. 3.7 The SAED of the (100) 3C-SiC/wafer observed along the [110] orientation of electron incidence

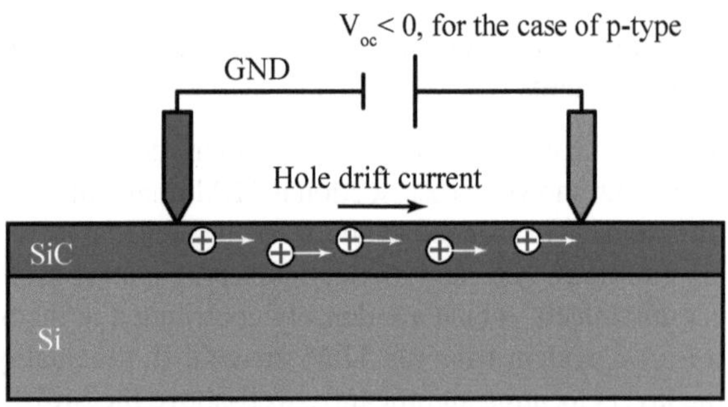

Fig. 3.8 A schematic sketch of the measurement using the hot probe technique. In the case of p-type semiconductor, since holes move from the hot probe towards the cold probe, a negative voltage is observed at the cold terminal with respect to the hot terminal

The semiconductor type (n-type and p-type) can be determined based on the sign of the voltage measured between the hot and cold probes. Additionally, the carrier concentration of the film can also be estimated from the measured voltage of the opened circuit $V_{oc} = P_{N,P}\Delta T$, where ΔT is the temperature different between the hot and cold probe, while $P_{N,P}$ (the subscriptions N and P denote n-type or p-type) is the thermoelectric power depending on the carrier concentration, according to [11]:

$$\begin{cases} P_N = -k_B/q(2.5 - s + ln(\frac{N_C}{n})) \\ P_P = +k_B/q(2.5 - s + ln(\frac{N_V}{p})) \end{cases} \quad (3.1)$$

where k_B, and q are the Boltzmann's constant, and the unit charge, respectively; N_C and N_V are the effective densities of state in the conduction and valence band, respectively; n and p are free electron and hole, respectively; and s is a factor depending on the free carrier scattering mechanism. The measured positive voltage at the cold probe in respect of the hot probe indicated that the 3C-SiC films were p-type. Additionally, the carrier concentrations were found to range from 5×10^{18} to 2×10^{19} cm^{-3}, according to Eq. 3.1.

3.2 Fabrication of 3C-SiC Piezoresistors

3C-SiC piezoresistors was fabricated using a two-masks photolithography process [4]. Starting from the epitaxial SiC/Si wafer (step 1), the SiC micro structures was patterned in an inductively coupled plasma (ICP) etcher through to the Si substrate (step 2) (Fig. 3.9). In the ICP etching step, a photoresist (AZ6612) with a thickness of 1 μm was used as the protect mask, while HCl and O_2 were utilized as the reactive gases with an etching rate of approximately 100 nm/min. Subsequently, a 100 nm aluminum film was deposited on the SiC/Si films using metal sputtering (step 3). Next, Al electrodes were wet-etched using Al etchant (HNO_3 : H_3PO_4 : CH_3COOH) (step 4). At each end of the piezoresistors, two Al electrode pads were created which allowed four-point measurements to eliminate contact resistance. Finally, the fabricated wafer was then diced into smaller strips with dimensions of 60 mm × 9 mm × 0.625 mm for the subsequent bending experiments (step 5). A detailed description of the fabrications process is presented in Appendix A.

Figure 3.10 shows a photograph of a SiC/Si beam used in the subsequent bending experiment, and a SEM image of a SiC resistor fabricated on the Si cantilever.

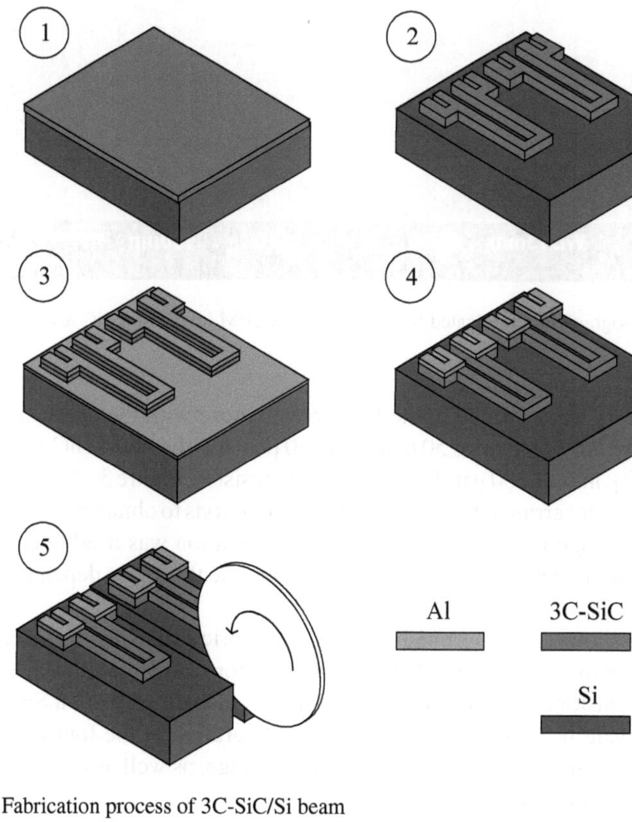

Fig. 3.9 Fabrication process of 3C-SiC/Si beam

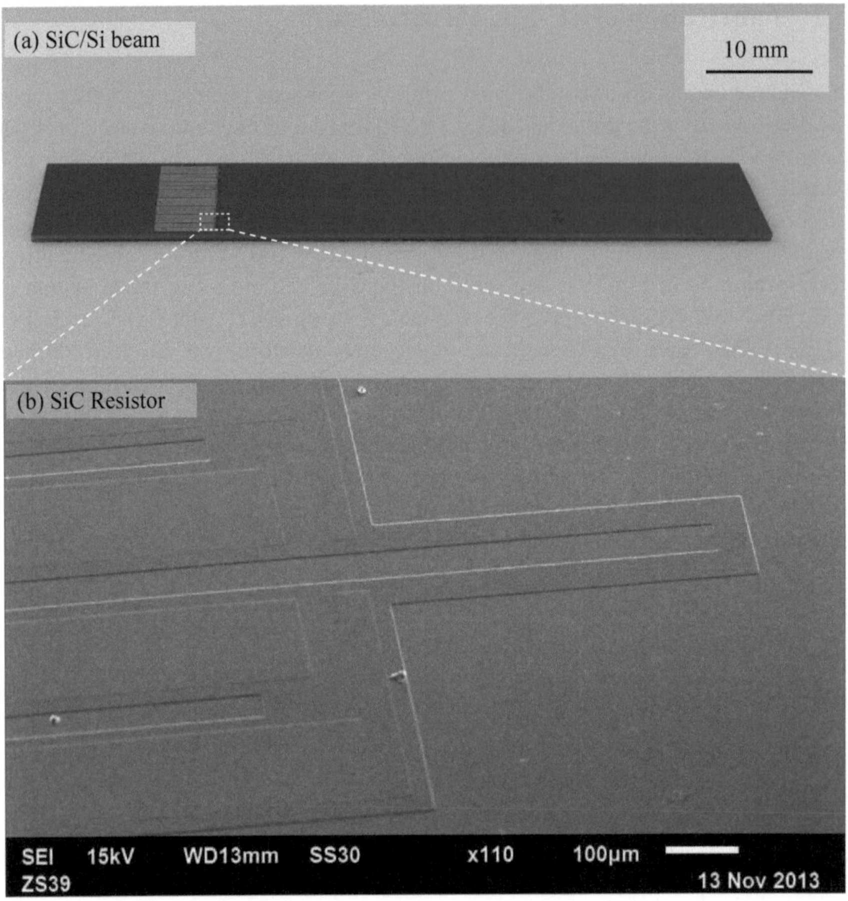

Fig. 3.10 **a** Photograph of a fabricated SiC/Si beam. **b** A SEM image of a SiC resistor

The SiC resistors were formed in the U-shape structure with the dimensions of the SiC resistors being 550 μm × 50 μm × 0.280 μm for the short SiC resistor, and 1050 μm × 50 μm × 0.280 μm for the long SiC resistor. Figure 3.11 shows the SiC resistors which were arranged in [110] and [1$\bar{1}$0] directions to obtain both longitudinal and transverse gauge factors. The U-shaped configuration was used to characterize the gauge factor, the piezoresistive coefficients, and the thickness dependence of the piezoresistance in p-type 3C-SiC.

Figure 3.12 shows a four-terminal SiC resistor fabricated using the same process as that of the above-mentioned U-shape SiC resistors. A Si cantilever with four-terminal SiC resistors fabricated on its surface was also diced from a SiC/Si wafer for the subsequent bending experiment. The configuration of the four-terminal SiC resistor, including the orientation of current and voltage, as well as the applied stress will be presented in Chap. 4.

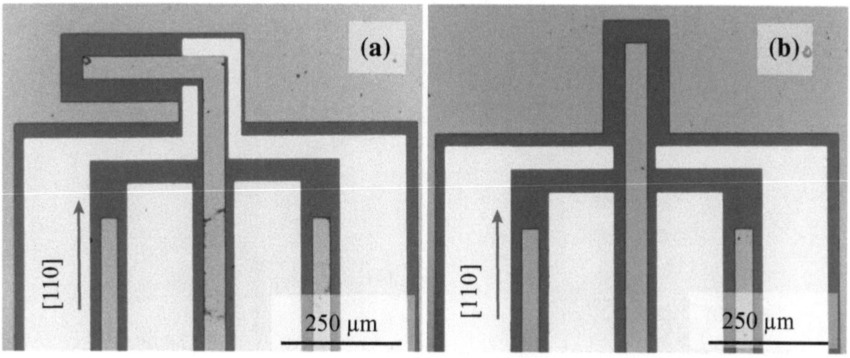

Fig. 3.11 **a** SiC resistor aligned in transverse [110] orientation. **b** SiC resistor aligned in longitudinal [110] orientation

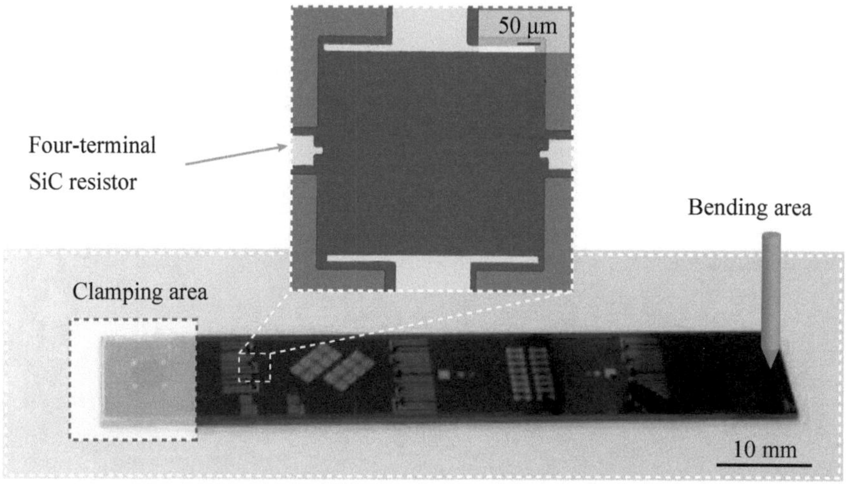

Fig. 3.12 Photograph of the SiC devices and a Si cantilever used to investigate the piezoresistive effect in p-type 3C-SiC four-terminal resistors

Furthermore, to investigate the piezoresistance at high temperatures, another configuration of SiC resistors was utilized, in which SiC bridges were released from the substrate, allowing the Joule heating effect. Additionally, to measure the temperature of the locally heated SiC bridges, SiC resistors were transferred onto an insulator to study its themoresistive effect. Figure 3.13 illustrates the fabrication of these SiC resistors.

Firstly, SiC resistors with Al electrodes were patterned using the same process as U-shape resistors as described in Fig. 3.9. Secondly, to create suspended SiC resistors, the Si substrate was under-etched by isotropic dry-etching [12], where XeF_2 was the etching gas (step 2). Afterward, samples with the suspended SiC bridge were inserted into a working chamber of a FIB machine, and a micro-probe was attached to the SiC

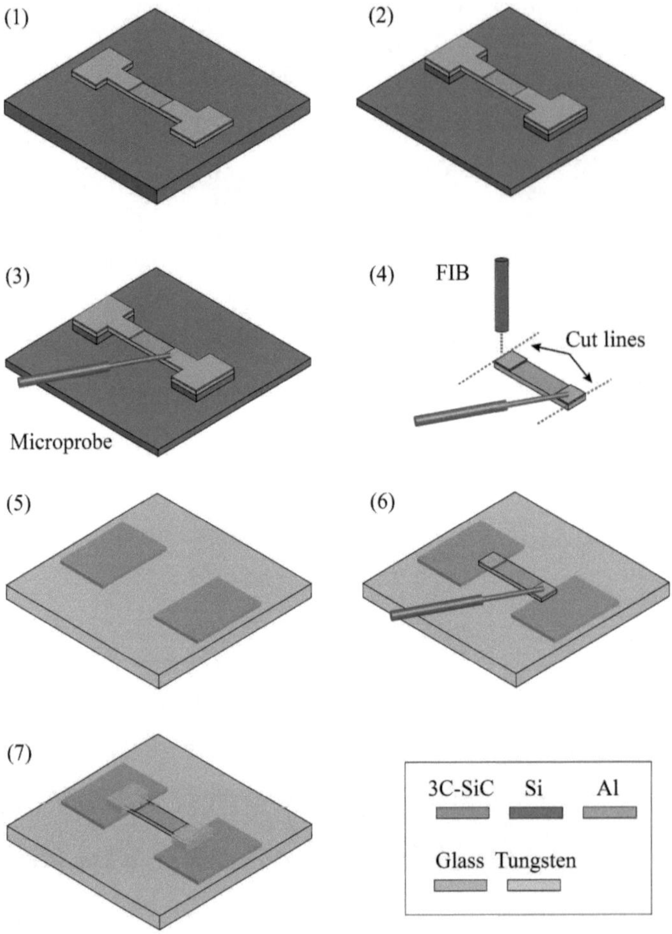

Fig. 3.13 Fabrication process of the suspended SiC bridges and the transferred SiC resistors for experiments at high temperatures

resistors (step 3). The two supporting hinges of the SiC resistors was then cut, using Ga^+ ions of the FIB (step 4). A glass wafer with aluminum electrodes deposited on it was also prepared (step 5), and the SiC resistors were then transferred onto the glass substrate [10, 13]. Finally, the micro-probe was removed, and subsequently tungsten was deposited to make the contact between the Al electrodes deposited on the glass substrate and the SiC resistors (step 7). Figure 3.14 shows photographs of a Si cantilever with released SiC resistors on its surface, and a SiC bridge transferred on a glass substrate.

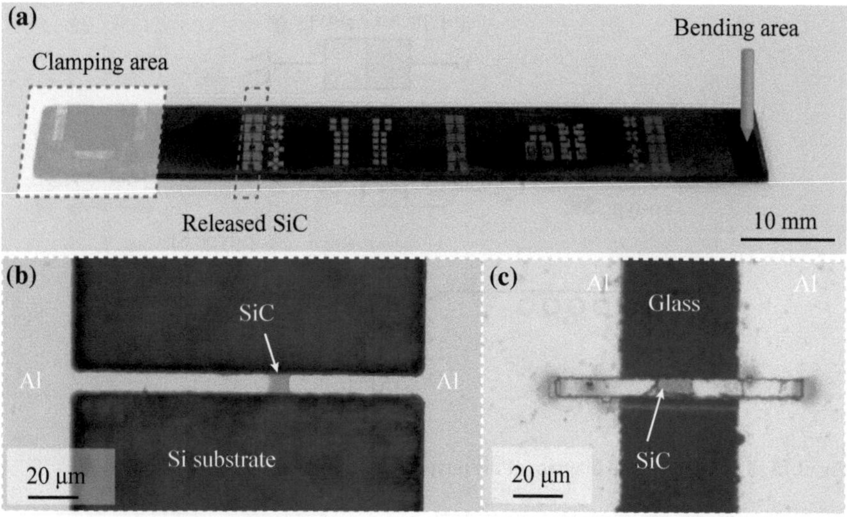

Fig. 3.14 **a** SiC/Si cantilever used for characterizing the piezoresistive effect at high temperature; **b** A released SiC bridge; **c** A transferred SiC resistor

3.3 Characterization of Ohmic Contact of SiC Resistors and Leakage Current to the Substrate

Since the as-grown p-type 3C-SiC films with an acceptor concentration of ($N_a = 5 \times 10^{18}\,cm^{-3}$) were deposited on low doped p-type Si with an acceptor concentration of ($N_a = 5 \times 10^{14}\,cm^{-3}$), the current leakage through the SiC/Si heterojunction was measured to investigate the influence of the Si substrate to the gauge factor of 3C-SiC. Figure 3.15 illustrates the energy diagram of the p-type SiC and p-type Si heterojunction. Accordingly, owing to the large discontinuity between the valence bands of 3C-SiC (Ev = 6.9 eV) and Si (Ev = 5.2 eV), a large potential height of 1.7 eV is built at the SiC/Si junction [14]. This large built-in potential barrier blocks holes in the valence bands from tunneling through the junction [15]. Moreover, the width of the depletion layer at the junction is increased when the SiC film is positively biased in respect of the Si layer.

The leakage current was then measured by sweeping voltages of SiC with respect to Si from −2 to 2 V using an HP 4145B Semiconductor Parameter Analyzer. Evidently, at a positive voltage, the leakage current was relatively small, being 2.5 nA at a supplied voltage of +0.5 V, Fig. 3.16a. Therefore, to measure the resistance of SiC under strains, a positive bias was applied to the SiC electrode, while the Si was grounded. In addition, to reduce the Joule heat effect, the current of the SiC resistors was set to 5 μA, at which the leakage current was only about 0.05% of the applied current, which can be neglected. Moreover, the I-V (current-voltage) curve of the SiC resistors exhibited good linearity indicating a good Ohmic contact between Al and SiC, Fig. 3.16b.

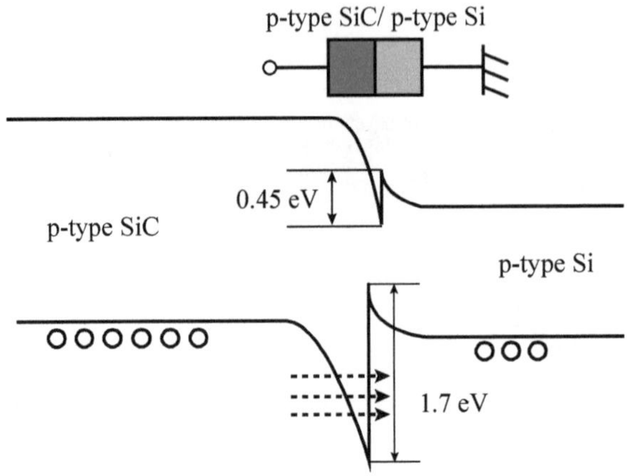

Fig. 3.15 The energy band diagram of the p type 3C-SiC/Si heterojunction

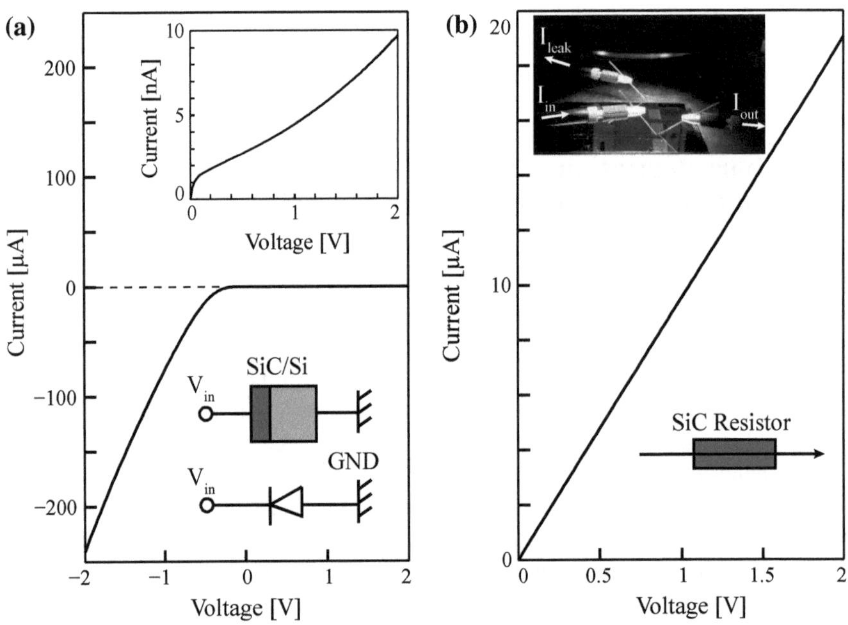

Fig. 3.16 Current-Voltage curves. **a** The current leakage to the Si substrate (*Inset* the current leakage at positive bias); **b** The current-voltage (I-V) curve of a SiC resistor

References

1. L. Wang, S. Dimitrijev, J. Han, A. Iacopi, L. Hold, P. Tanner, H.B. Harrison, Growth of 3C-SiC on 150-mm Si(100) substrates by alternating supply epitaxy at 1000°C. Thin Solid Films **519**, 6443–6446 (2011)
2. L. Wang, S. Dimitrijev, J. Han et al., Demonstration of p-type 3C-SiC grown on 150 mm Si(100) substrates by atomic- layer epitaxy at 1000 °C. J. Crystal Growth **329**, 67–70 (2011)
3. L. Wang, S. Dimitrijev, G. Walker et al., Color chart for SiC thin films grown on Si substrate. Mater. Sci. Forum **740–742**, 279–282 (2013)
4. H.-P. Phan, P. Tanner, D.V. Dao, N.-T. Nguyen, L. Wang, Y. Zhu, S. Dimitrijev, Piezoresistive effect of p-type single crystalline 3C-SiC. IEEE Electron Devices Lett. **35**(3), 399–401 (2014)
5. R.J. Iwanowski, K. Fronc, W. Paszkowicz, M. Heinonen, XPS and XRD study of crystalline 3C-SiC grown by sublimation method. J. Alloys Compd. **286**(1), 143–147 (1999)
6. F. Smith (ed.), *Industrial Applications of X-Ray Diffraction* (CRC Press, Boca Raton, 1999)
7. H.-P. Phan, D.V. Dao, P. Tanner, N.-T. Nguyen, L. Wang, Y. Zhu, S. Dimitrijev, Fundamental piezoresistive coefficient of p-type single crystalline 3C-SiC. Appl. Phys. Lett. **104**, 111905 (2014)
8. M. Portail, M. Zielinski, T. Chassagne, S. Roy, M. Nemoz, Comparative study of the role of the nucleation stage on the final crystalline quality of (111) and (100) silicon carbide films deposited on silicon substrates. J. Appl. Phys. **105**, 083505 (2009)
9. H.-P. Phan, D.V. Dao, P. Tanner, J. Han, N.-T. Nguyen, S. Dimitrijev, G. Walker, L. Wang, Y. Zhu, Thickness dependence of the piezoresistive effect in p-type single crystalline 3C-SiC film. J. Mater. Chem. C **2**, 7176–7179 (2014)
10. H.-P. Phan, T. Kozeki, T. Dinh, T. Fujii, A. Qamar, Y. Zhu, T. Namazu, N.-T. Nguyen, D.V. Dao, Piezoresistive effect of p-type silicon nanowires fabricated by a top-down process using FIB implantation and wet etching. RSC Adv. **5**, 82121–82126 (2015)
11. P. Tanner, L. Wang, S. Dimitrijev, J. Han, A. Iacopi, L. Hold, G. Walker, Novel electrical characterization of thin 3C-SiC films on Si substrates. Sci. Adv. Mater. **6**(7), 1542–1547 (2014)
12. A.R. Kermany, G. Brawley, N. Mishra, E. Sheridan, W.P. Bowen, F. Iacopi, Microresonators with Q-factors over a million from highly stressed epitaxial silicon carbide on silicon. Appl. Phys. Lett. **104**(8), 081901 (2014)
13. T. Fujii, T. Namazu, K. Sudoh, S. Sakakihara, S. Inoue, Focused ion beam induced surface damage effect on the mechanical properties of silicon nanowires. J. Eng. Mater. Technol. **135**(4), 041002 (2013)
14. P. Tanner, S. Dimitrijev, H.B. Harrison, Current mechanisms in n-SiC/p-Si heterojunctions, in *IEEE Conference on Optoelectronic and Microelectronic Materials and Devices, COMMAD* (2008), pp. 41–43
15. A. Qamar, P. Tanner, D.V. Dao, H.P. Phan, T. Dinh, Electrical properties of p-type 3C-SiC/Si heterojunction diode under mechanical stress. IEEE Electron Device Lett. **35**(12), 1293–1295 (2015)

References

Chapter 4
Characterization of the Piezoresistive Effect in p-Type Single Crystalline 3C-SiC

This chapter presents experimental work on the piezoresistance of p-type 3C-SiC two-terminal and four-terminal resistors. Firstly, the experimental method to induce strain into SiC piezoresistors is discussed. Secondly, the gauge factor, the orientation dependence, the thickness dependence, as well as the piezoresistance at high temperature in p-type 3C-SiC are presented. Finally, the piezoresistive effect of four-terminal resistors is described at the end of this chapter.

4.1 Measurement Method

The bending experiment method was employed to investigate the piezoresistive effect in p-type 3C-SiC, as illustrated in Fig. 4.1a. In this experimental setup, a SiC/Si cantilever was fixed at one end, while the other end was deflected by applying load. Consequently, the deflection of the SiC/Si cantilever will induce a strain into the SiC resistors lying on the Si substrate. To induce a large strain into SiC which could ease the measurement of the resistance change, all SiC resistors were formed at the vicinity of the fixed area of the Si beam. The applied strain was then estimated as follows.

In the case of a uniform cantilever where the piezoresistor and the substrate are made of the same material, as presented in Fig. 4.1b, the strain induced into the piezoresistor is almost equal to that applied to the surface of the cantilever, which is calculated using the equation below:

$$\varepsilon = Mt/(EI) \tag{4.1}$$

where M is the bending moment, E is the Young's modulus of the uniform cantilever, I is the area moments of inertia, and t is the thickness of the beam, respectively. However, because SiC and Si have different Young's moduli, the bending model

H.-P. Phan, *Piezoresistive Effect of p-Type Single Crystalline 3C-SiC*, Springer Theses, DOI 10.1007/978-3-319-55544-7_4

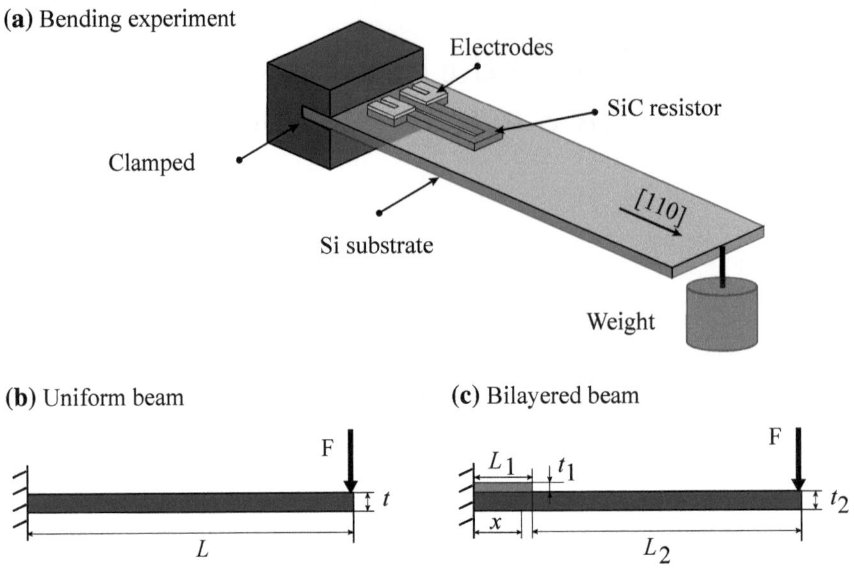

Fig. 4.1 **a** The bending experiment; **b** An uniform beam model; **c** A bi-layer beam model

of a bi-layered beam was utilized to estimate the strain applied to the SiC layer ($\varepsilon_{\mathrm{SiC}}$). This bi-layered model has been widely applied in piezoelectric multilayer beam-bending-actuators, where the bimorph (two active piezoelectric layers) and the trimorph (three active piezoelectric layers) are widely utilized [1]. Figure 4.1c illustrates the concept of the SiC and Si bi-layers, where the lateral strain of a SiC piezoresistor can be driven from the mechanical properties and dimensions of the bi-layers [2]:

$$\varepsilon_{\mathrm{SiC}}(x) = -\frac{F}{wD_1}(L_1 + L_2 - x)t_n \tag{4.2}$$

where F is the applied force; t_n is the distance from neutral axis to the SiC layer; and parameters w, L_1, L_2 are dimensions as described in Fig. 4.1c. The bending modulus per unit width D_1 is deduced from [3]:

$$D_1 = \frac{E_1^2 t_1^4 + E_2^2 t_2^4 + 2E_1 E_2 t_1 t_2 (2t_1^2 + 2t_2^2 + 3t_1 t_2)}{12(E_1 t_1 + E_2 t_2)} \tag{4.3}$$

where E_1 is the Young's modulus of SiC (330 GPa) [4], and E_2 is the Young's modulus of Si (130 GPa in [100] direction and 169 GPa in [110] direction). As the thickness of SiC film employed in this work was below 1 μm, which was less than 0.2% of the Si layer (625 μm), the strain of the SiC piezoresistors calculated using Eq. 4.2 was almost equal to that estimated using Eq. 4.1.

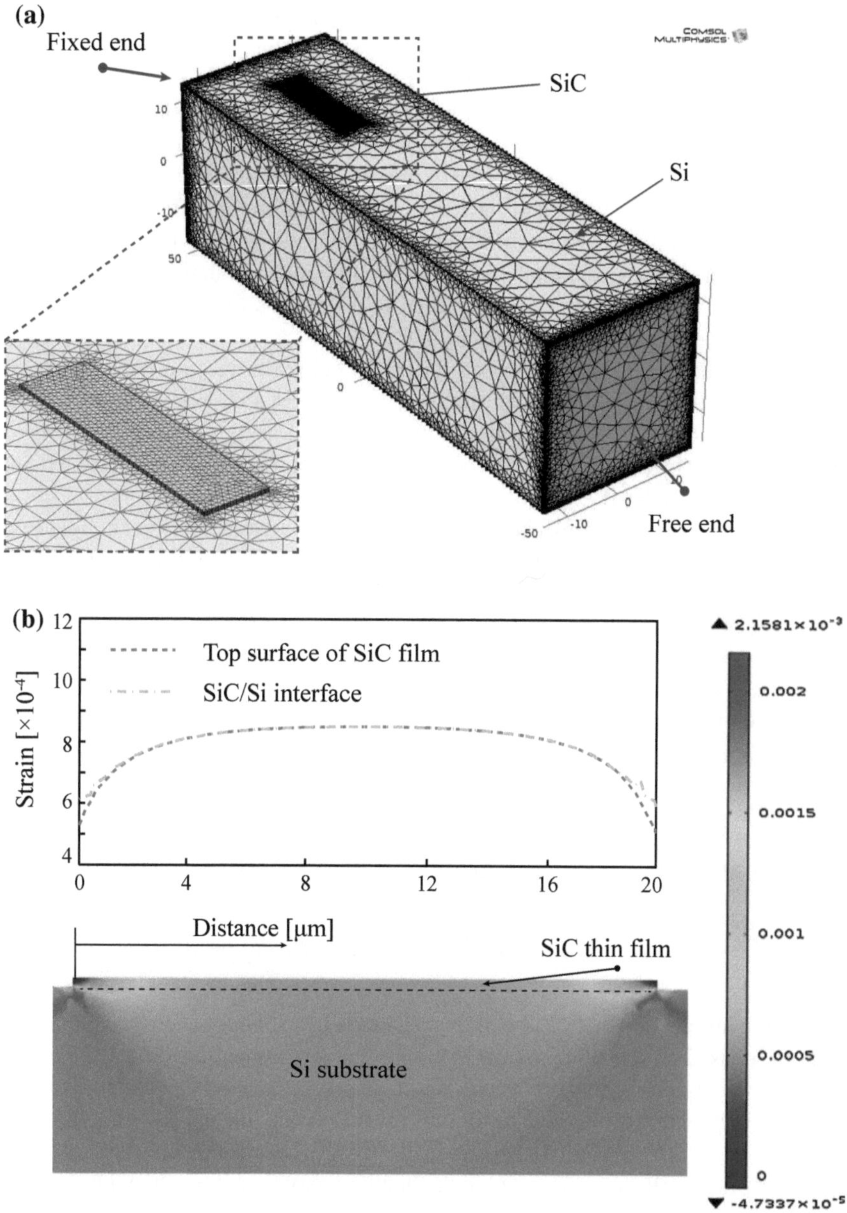

Fig. 4.2 **a** Simulation of strain applied to a SiC/Si beam; **b** Strain distribution along a SiC resistor (upper part)

Simulation of the bending experiment was also conduct using finite element analysis (FEA) to estimate the applied strain and to compare with the numerically calculated results. Figure 4.2 shows the strain distribution in a 280 nm thin SiC film simulated using ™COMSOL Multiphysics. The results show that the strain applied to the Si surface was transmitted by at least 98% to the SiC layer. Additionally, the strain was almost uniformly distributed along the SiC nano thin film and only a smaller strain was observed at the top-edge of the SiC film. However, since the length of this edge part is much smaller than the total length of the SiC resistor, the influence of this small strain at this area of the SiC piezoresistor is insignificant. The same simulation studies were also carried out on SiC films with different thicknesses ranging from 80 nm to 1 μm, exhibiting the same trend as shown in Fig. 4.2. The above-mentioned numerical analysis and computational simulations indicated that for the sake of simplicity, the strain applied to a thin SiC film grown on a bulk Si substrate can be approximated using Eq. 4.1.

This bending experiment was then utilized to characterize the piezoresistance of two-terminal and four-terminal single crystalline 3C-SiC resistors, as well as nanocrystalline 3C-SiC, which are presented in the following sections. The error of the strain caused by the fabrication process (e.g. misalignment) is also discussed and presented in Appendix B.

Figure 4.3 shows the setup of bending experiments. To avoid the effect of light (e.g. photon absorption of SiC and Si) all experiments were conducted in a dark condition. Additionally, different weights were applied to the SiC/Si cantilever, using a linear motor. The applied loads were switched to on– and off–states, by actuating the motor from its highest to lowest positions, respectively. To stabilize the experiments, the setup was covered in an enclosed chamber.

A list of electrical devices used for measuring the piezoresistive effect in SiC are listed in Table A.2, Appendix A.

4.2 Measurement Results at Room Temperature

4.2.1 The Gauge Factors of Two Terminal p-Type 3C-SiC

As aforementioned in Chap. 2, the piezoresistance in a semiconductor is commonly quantified based on its gauge factor. Utilizing the above-mentioned bending experiment, this section presents the gauge factor of p-type 3C-SiC calculated from the relationship between the change in the SiC resistance and the applied strains.

The U-shape SiC resistors with different dimensions aligned in longitudinal [110], and transverse [110] were subjected to the bending experiment, as shown in Fig. 4.4. The U-shape was selected as it is a common configuration used in several piezoresistive applications such as AFM cantilevers, tactile sensors, and acceleration sensors [5, 6]. The thickness of the epitaxial 3C-SiC films used in this experiment was 280 nm. Furthermore, the dimensions of the Si beam were 60 mm in length, 9 mm in width, and 625 μm in thickness, respectively. Consequently, for applying loads varying from 0 to approximately 2 N, the strains applied to SiC piezoresistors ranged from 0 to 1000 ppm, using Eq. 4.1.

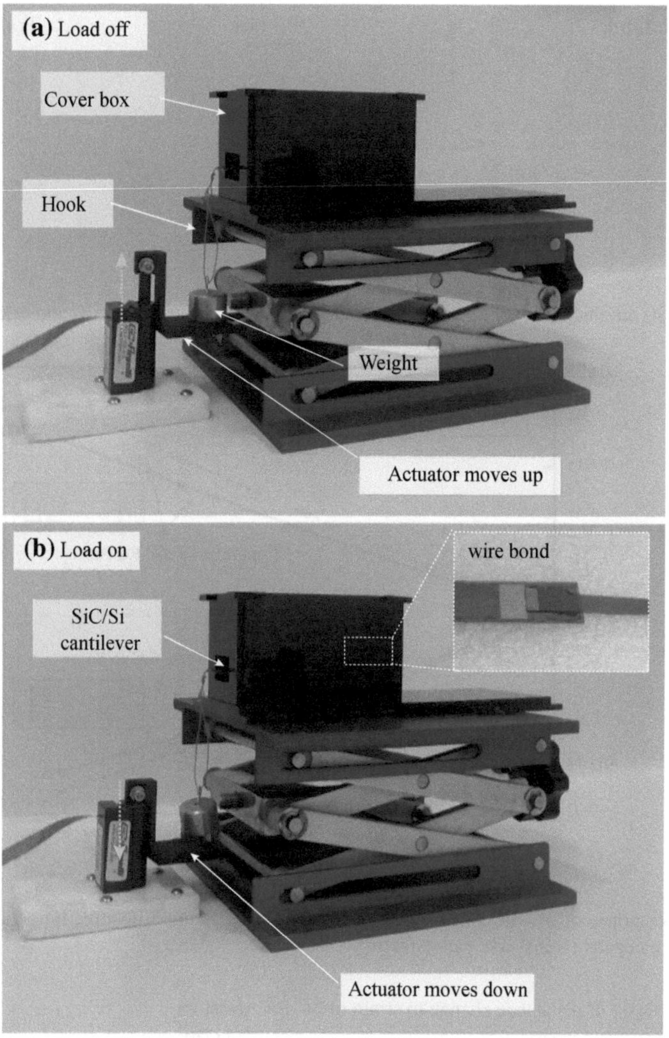

Fig. 4.3 Photographs of bending experiments in which an external force was applied by hanging a weight on a hook attached at the free end of a SiC/Si cantilever. **a** Load off; **b** Load on (*inset* the electrical contacts were made using wire bonding.)

The resistance change of SiC under strain was measured using the 4 point probe measurement where the current was applied at the outer electrodes, while the sensing voltage was measured at the inner electrodes, using an ™Agilent 34410A Multimeter. Figure 4.5 shows the relative resistance change ($\Delta R/R$) of p-type SiC against applied tensile strains (ε). Evidently, for SiC resistors aligned in the longitudinal direction of the Si beam, the resistance increased with increasing the tensile strain. However, for SiC resistors aligned in the transverse direction, the resistance decreased when

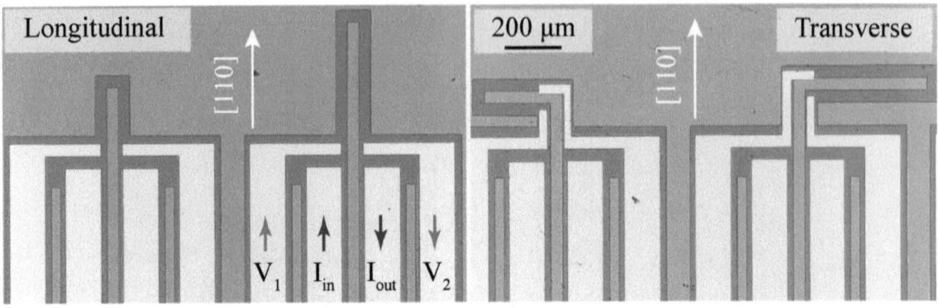

Fig. 4.4 280 nm thick p-type 3C-SiC resistors designed with different lengths and orientations

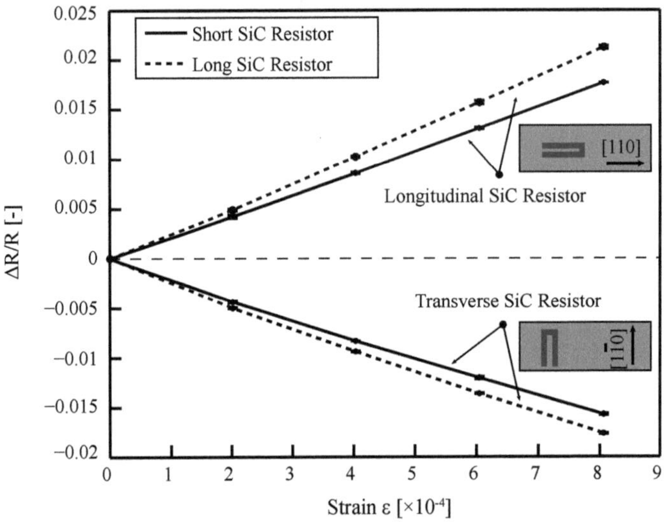

Fig. 4.5 Response of 3C-SiC piezoresistor under strain, showing a positive longitudinal gauge factor, and a negative transverse gauge factor

Table 4.1 Ratio of resistance change to strain of 3C-SiC resistors

Orientation	[110]		[1$\bar{1}$0]	
Length(μm)	1050	550	1050	550
$(\Delta R/R)/\varepsilon$	25.8	21.6	−22.6	−20.0

the tensile strain increased. Additionally, for all SiC piezoresistors, the resistances exhibited good linearity with the applied strains. Table 4.1 shows the ratios of the resistance change $\Delta R/R$ to the normal strains ε in [110] direction of SiC resistors designed with different lengths and orientations.

Evidently, for both longitudinal and transverse resistors, the long SiC resistors show a larger ratio of the relative resistance change per applied strain ($|(\Delta R/R)/\varepsilon|$) than short resistors. This result is caused by the configuration of U-shape resistors

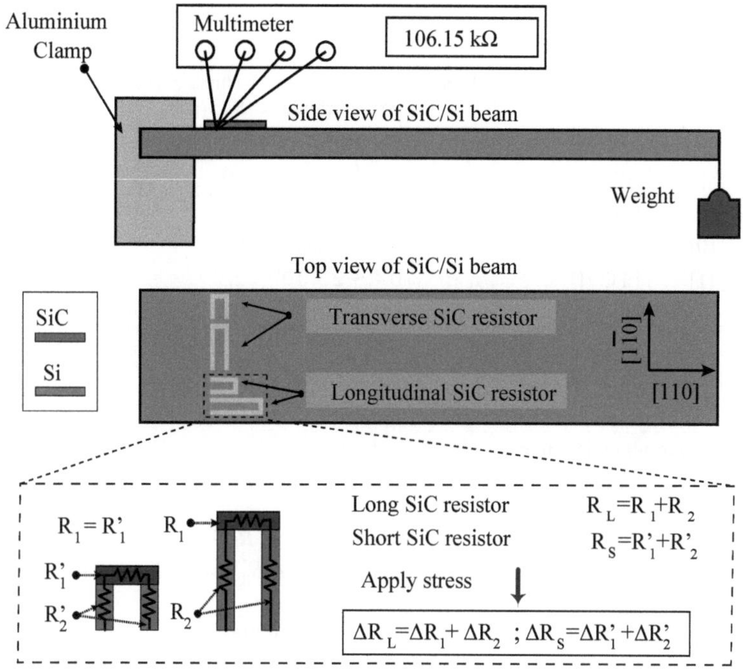

Fig. 4.6 Experimental setup in which the SiC resistors were aligned in [110] and [1$\bar{1}$0] directions and each SiC resistor consisted of a longitudinal section R_2 and a transverse section R_1

which consists of both longitudinal and transverse sections, having different dimensions. Therefore, the following method was developed to estimate the longitudinal and transverse gauge factors.

Considering a pair of long and short longitudinal SiC resistors R_L and R_S aligned in [110] orientation as shown Fig. 4.6. Each resistor is configured by two parts R_1 (or R'_1) and R_2 (or R'_2 for the short resistor) aligned in transverse and longitudinal directions. Under a [110] tensile strain, the resistance changes ΔR_1 of section R_1 and $\Delta R'_1$ of section R'_1 were in accordance with the transverse gauge factor, whereas the resistance changes ΔR_2 of section R_2 and $\Delta R'_2$ of section R'_2 correspond to the longitudinal gauge factor. Because the long and short SiC resistors were formed at almost the same position, the strains indued into these two resistors were considered to be equal. Consequently, the longitudinal gauge factor (GF_L) of p-type 3C-SiC in [110] orientation can be obtained using the equation below:

$$GF_L = \frac{1}{\varepsilon}\frac{\Delta R_2}{R_2} = \frac{1}{\varepsilon}\frac{\Delta R'_2}{R'_2} = \frac{1}{\varepsilon}\frac{\Delta R_2 - \Delta R'_2}{R_2 - R'_2} \tag{4.4}$$

Furthermore, since the transverse parts R_1 and R'_1 of the long and short SiC resistors were designed with the same dimensions ($R_1 = R'_1$ and $\Delta R_1 = \Delta R'_1$), the following relationships are established:

Table 4.2 A comparison between the gauge factor found in this work and that of SiC films reported in the literature

Polytype	Growing process	Type	Carriers concentration	Thickness of SiC (μm)	Gauge factor
3C-SiC[a]	LPCVD	p	5×10^{18}	0.28	30.3
3C-SiC [8]	APCVD	n	$\sim 10^{18}$	10	−31.8
3C-SiC [9]	HMCVD	n	$\sim 10^{18}$	2	−27
3C-SiC [10]	APCVD	n	Unintentional	0.5	−18.8
3C-SiC [11]	LPCVD	n	$0.4 \sim 2 \times 10^{17}$	2.3	−24.8
				0.2	3
4H-SiC [12]	–	n	1.5×10^{19}	1	20.8
6H-SiC [13]	–	n	3.8×10^{18}	2	−29.4
6H-SiC [14]	–	p	2×10^{19}	2	27

[a] Indicated the result in this study

$$\begin{cases} R_2 - R_2' = (R_2 + R_1) - (R_2' + R_1') = R_L - R_S \\ \Delta R_2 - \Delta R_2' = (\Delta R_2 + \Delta R_1) - (\Delta R_2' + \Delta R_1') = \Delta R_L - \Delta R_S \end{cases} \tag{4.5}$$

Consequently, the longitudinal GF_L in [110] orientation is:

$$GF_L = [(\Delta R_L - \Delta R_S)/(R_L - R_S)]/\varepsilon \tag{4.6}$$

Based on the above equation, the longitudinal (GF_L) and transverse (GF_T) gauge factors in [110] orientation of p-type single crystalline 3C-SiC were found to be 30.3 and −25.1, respectively, which are considerably larger than that of the p-type poly 3C-SiC reported by Homma et al. [7].

Table 4.2 shows the comparison between the gauge factors of p-type 3C-SiC presented in this work with other single crystal SiC films. Evidently, the gauge factor of the p-type 3C-SiC presented in this work is comparable to the p-type 6H-SiC reported in [14], with a gauge factor of 27. In addition, the gauge factor of p-type 3C-SiC thin film is larger than that of n-type 3C-SiC deposited by APCVD with the same range of carrier concentration [8].

The large gauge factor of 30 indicates the potential of p-type single crystal 3C-SiC for MEMS mechanical sensors. Figure 4.7a illustrates a schematic sketch of p-type 3C-SiC strain sensors. The strain of an object can be monitored from the resistance change of a SiC piezoresistor mounted on its surface. In fact, the SiC on Si surface structure used in this study can also be considered as a strain sensor, since it has the capability of detecting strains applied to the Si beam. In practical use, the SiC resistance change can be converted to a voltage signal to ease the measurement, as well as data recording, using a Wheatstone bridge circuit and a signal amplifier, Fig. 4.7b. Given that V_i is applied voltage, and G is the amplified gain of the Wheatstone bridge, the the output voltage (V_o) of a SiC strain sensor is:

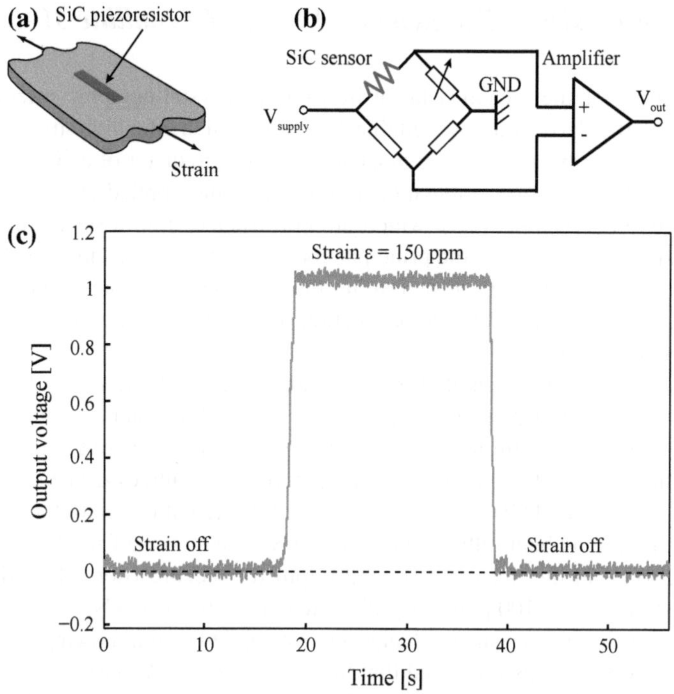

Fig. 4.7 Demonstration of a strain sensors using the piezoresistive effect in p-type 3C-SiC. **a** A schematic sketch of a strain sensor attached on the surface of an object; **b** The Wheatstone bridge and amplifier circuit used to convert the resistance change to voltage signals; **c** The output of a SiC strain sensor subjected to a tensile strain of 150 ppm

$$V_o = \frac{V_i}{4} \times \frac{\Delta R}{R} \times G \tag{4.7}$$

Subsequently, from the variation of the output voltage (V_o) and the gauge factor of SiC piezoresistors (GF), the strain impacting on the surface where SiC is attached onto is estimated using the equation below:

$$\varepsilon = \frac{\Delta R/R}{GF} = 4 \times \frac{V_o}{V_i \times G \times GF} \tag{4.8}$$

Utilizing the Wheatstone bridge and a voltage amplifier (AD623) with a gain of 1000 times, a measurement of strains induced into the Si surface was demonstrated, as shown in Fig. 4.7c. The output of the sensor measured using an oscilloscope (TMAgilent MSO-X 3104A) showed good reproducibility and stability when subjected to a strain of 150 ppm for several cycles.

4.2.2 Piezoresistive Coefficients of Single Crystalline 3C-SiC

Understanding of the piezoresistance in arbitrary crystallographic orientations is important, particularly to choose suitable directions which offer high strain/stress sensitivity. Additionally, as described in Chap. 3, the gauge factor of a 3C-SiC piezoresistor aligned in arbitrary orientation is related to the applied stress through the piezoresistive coefficients (π'_{ij}), which are constructed from a set of fundamental piezoresistive coefficients in the principal coordinate (π_{11}, π_{12}, and π_{44}). Therefore, the aims of this section are to determine these fundamental piezoresistive coefficients, as well as to elucidate the orientation dependence of the piezoresistance of p-type single 3C-SiC.

There are three fundamental piezoresistive coefficients (π_{11}, π_{12}, and π_{44}); therefore, at least three independent equations are required to obtain these coefficients. A common way to determine these parameters is to characterize the piezoresistive effect in three different orientations. Consequently, SiC samples aligned in longitudinal [110], longitudinal [100], and transverse [100] orientations were fabricated with their configurations being illustrated in Fig. 4.8. The strains of the SiC layer were estimated to be in the ranges of $0 \sim 820$ ppm for piezoresistors aligned in [110] orientation, and $0 \sim 1100$ ppm for [100] orientation, respectively.

Figure 4.9 plots the resistance change against applied strain, showing a good linear behavior. The results also revealed that the resistance of SiC piezoresistors aligned in longitudinal [110] and longitudinal [100] orientations increased with increasing tensile strain, while the piezoresistors aligned in transverse [100] direction reduced their resistance when the tensile strain was increased. It is also evident that SiC piezoresistors aligned in [110] direction were more sensitive than those aligned in [100] orientation, implying the orientation dependent property of the piezoresistance in the p-type 3C-SiC crystal. Subsequently, the gauge factors in different orientations were calculated using the method which was presented in Sect. 4.2.1. Accordingly, positive gauge factors of +5.0 and +30.3 were observed in [100] and [110] longitu-

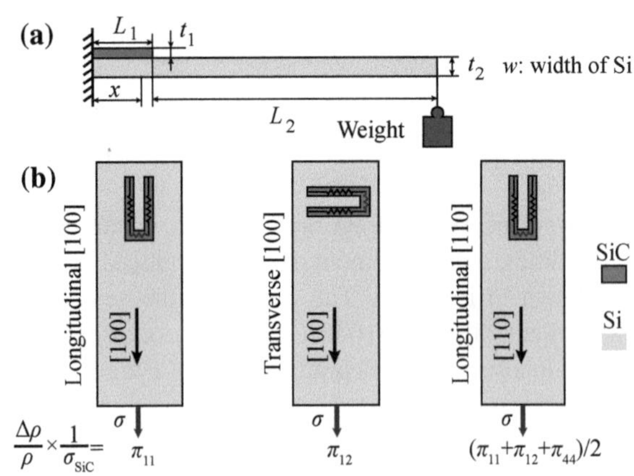

Fig. 4.8 A schematic diagram showing the bending experiment for measuring the gauge factors. **a** Side view; **b** Top view of cantilever

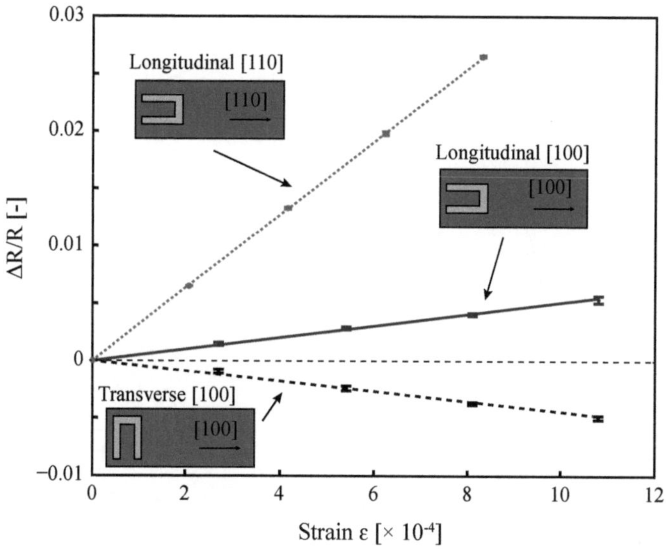

Fig. 4.9 The resistance change of p-type SiC aligned in different orientations under strain

dinal orientations, respectively. The transverse gauge factor in [100] orientation, on the other hand, possessed a negative value of −4.6.

Subsequently, the gauge factors measured above were employed to calculate the fundamental piezoresistive coefficients. The change of the resistivity ($\Delta\rho/\rho$) of SiC under strain (ε) is [15]:

$$\Delta\rho/\rho = GF \times \varepsilon = \pi_l\sigma_l + \pi_t\sigma_t + \pi_s\sigma_s \tag{4.9}$$

where σ_l, σ_t, and σ_s are longitudinal, transverse and shear stresses, respectively, and π_l, π_t, and π_s are longitudinal, transverse and shear piezoresistive coefficients, respectively. Let θ be the angle between the longitudinal axis of the SiC resistor and [100] orientation in (100) plane. π_l, π_t, and π_s are obtained from the fundamental piezoresistive coefficients as [15]:

$$\begin{cases} \pi_l = \pi_{11} - \frac{1}{2}(\pi_{11} - \pi_{12} - \pi_{44})\sin^2 2\theta \\ \pi_t = \pi_{12} + \frac{1}{2}(\pi_{11} - \pi_{12} - \pi_{44})\sin^2 2\theta \\ \pi_s = -\frac{1}{2}(\pi_{11} - \pi_{12} - \pi_{44})\sin 4\theta \end{cases} \tag{4.10}$$

From Eqs. 4.9 and 4.10, and the gauge factors measured in [100] ($\theta = 0°$) and [110] ($\theta = 45°$) orientations, the relationship between the measured resistivity change and the piezoresistive coefficients is:

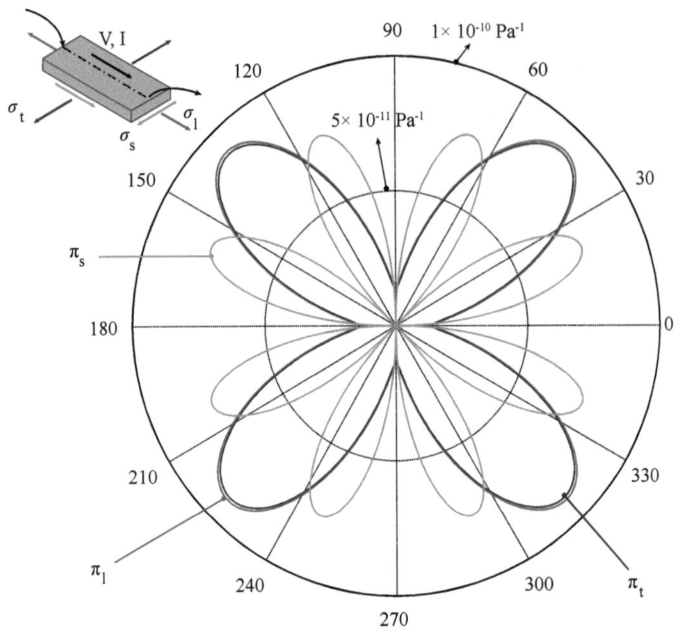

Fig. 4.10 Piezoresistive coefficients π_l, π_t, and π_s in (100) plane of p-type single crystalline 3C-SiC (In some literature [15], π_l, π_t, and π_s are denoted as π'_{11}, π'_{12} and π'_{16})

$$\begin{cases} \frac{\Delta\rho}{\rho}_{L,[100]} = \pi_{11}\varepsilon_{L,[100]} \\ \frac{\Delta\rho}{\rho}_{T,[100]} = \pi_{11}\varepsilon_{T,[100]} \\ \frac{\Delta\rho}{\rho}_{L,[110]} = \frac{1}{2}(\pi_{11} + \pi_{12} + \pi_{44})\varepsilon_{L,[110]} \end{cases} \tag{4.11}$$

The fundamental piezoresistive coefficient π_{11}, π_{12} and π_{44} were calculated to be 1.5×10^{-11} Pa^{-1}, -1.4×10^{-11} Pa^{-1}, 18.1×10^{-11} Pa^{-1}, respectively. This result indicates the similarity between the piezoresistance of p-type single 3C-SiC and that of p-type single Si, i.e. the shear piezoresistive coefficient (π_{44}) is much larger than normal piezoresistive coefficients (π_{11} and π_{12}).

As presented in Chap. 2, since 3C-SiC has the cubic crystal structure like Si, the model of strain effect in cubic semiconductors can be employed to explain the piezoresistance of single crystalline 3C-SiC [16]. The piezoresistive coefficients π_{11}, π_{12} are in accordance with the uniaxial stresses applied in [100] direction, where the band warping is considerably negligible, leading to a small gauge factor. On the other hand, the piezoresistive coefficients π_{44} corresponds to the uniaxial stress in [110] direction, where the energy surface of the top valance bands is significantly warped, resulting in the change of carrier concentrations in the heavy hole and light hole bands. This also leads to the modification of the hole effective mass in these two bands. Consequently, it is considerable that the uniaxial stress in [110] orientation exhibits more profound piezoresistance than the uniaxial stress in [100] orientation.

Piezoresistive coefficients in other orientations can also be estimated based on the fundamental coefficients using the transformation equations described in Chap. 2. Figure 4.10 shows the longitudinal (π_l), transverse (π_t) and shear (π_s) piezoresistive coefficients of an arbitrary orientation in 3C-SiC (100) plane. Evidently, the piezoresistance of p-type 3C-SiC exhibits an anisotropic property. As such, the transverse and longitudinal coefficients reach maximum values when the p-type SiC piezoresistors aligns in [110] orientation. Therefore, to obtain highly sensitive p-type 3C-SiC mechanical sensors (e.g. strain and pressure sensors), this crystal orientation should be selected. A demonstration of the estimation of output of p-type 3C-SiC piezoresistors under strain using these coefficients is presented in Appendix C.

4.2.3 Thickness Dependence of the Piezoresistive Effect of p-Type 3C-SiC Nano Thin Films

Recently, the effect of strain on the nano-scaled semiconductors is of great interest, in consideration of the exciting properties of low dimension structures such as nanotubes, nanowires, and 2D materials. The giant piezoresistance reported in Si nanowires [17] and Si nano thin film [18] with the piezoresistive coefficient of $-3350 \times 10^{-11}\text{Pa}^{-1}$ and $440 \times 10^{-11}\text{Pa}^{-1}$, respectively has been a motivation for investigation into the piezoresistance of nanoscale SiC. To date, there have been a few studies on SiC nanowires, which were fabricated using bottom up growth. These nanowires with diameters of 150 nm exhibited comparable piezoresistive coefficients with Si [19–21]. Compared to the bottom up method, the top down method has several advantages such as the compatibility with the conventional micromachining and packaging for MEMS devices. This section presents an experimental study on the thickness dependence of the piezoresistance of the p-type single crystal 3C-SiC nano thin films.

3C-SiC films with the thickness ranged from 80 nm to 1 μm were fabricated using the LPCVD process presented in Chap. 3. The thicknesses of these 3C-SiC films were controlled by varying the number of growth cycles. U-shaped SiC resistors with different thickness were then fabricated using the photolithography process. Since the 3C-SiC films had different thicknesses, these SiC resistors reflected different colors under an optical microscope, as shown in Fig. 4.11a–e. Additionally, based on the dependence on crystal orientation of the piezoresistance described in the previous section, to investigate the influence of the thickness of SiC films, SiC resistors were aligned in [110] orientation, since it offers the most significant piezoresistive effect in (100) plane of p-type 3C-SiC.

Table 4.3 shows the data obtained from the hot probe measurement. Accordingly, all as-grown SiC films were p-type, and had the same range of carrier concentration of approximately $1.4 \sim 10 \times 10^{18}\text{cm}^{-3}$.

Utilizing the bending experiment, the piezoresistive of p-type single crystalline 3C-SiC films with different thickness was investigated. Figure 4.12 plots the resis-

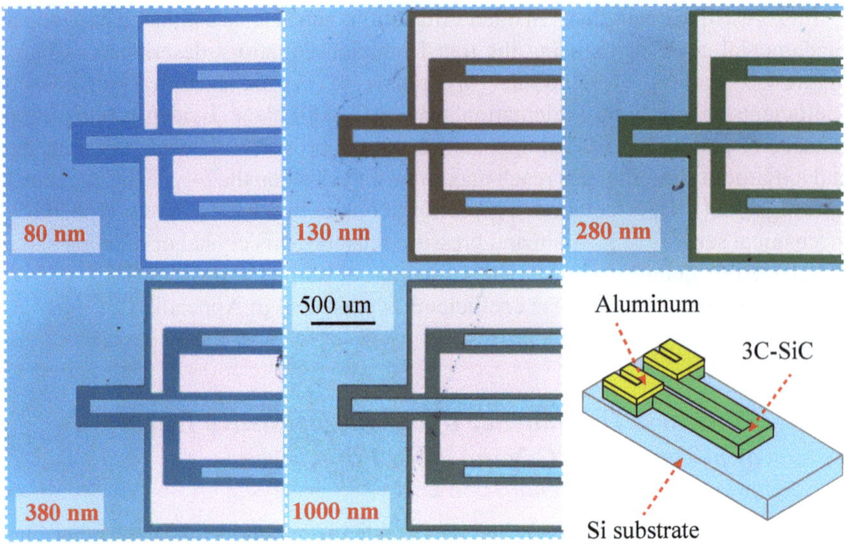

Fig. 4.11 From **a** to **e** Photograph of SiC resistors on Si substrate in which different thickness reflects different color. **f** Bending experiment (color figure online)

Table 4.3 Hot probe measurement of SiC films with different thickness

Film thickness (nm)	V_{oc} (mV)	I_{sc} n(A)	Carrier concentration N (cm^{-3})
1000	9.8	−1940	$1.4 \sim 6.2 \times 10^{18}$
380	8.5	−1057	$2.5 \sim 10 \times 10^{18}$
280	9.6	−670	$1.5 \sim 6.8 \times 10^{18}$
130	9.4	−480	$1.5 \sim 7.0 \times 10^{18}$
80	9.9	−75	$1.3 \sim 6.0 \times 10^{18}$

V_{oc}: the open circuit voltage; I_{sc}: the short circuit current [22]

tance of SiC films with different thickness against external mechanical strains. Evidently, for all SiC piezoresistors aligned in longitudinal [110] orientation, the resistances increased when increasing the tensile strain, indicating positive gauge factors. Furthermore, by using the calculation method reported in Sect. 4.2.1, the gauge factor of the 80, 130, 280, 380 nm and 1 μm films were found to be 20.5, 26.1, 30.3, 30.4 and 31.1, respectively. Evidently, no giant piezoresistance was observed in all as-grown nano thin SiC films. Additionally, it is evident that the gauge factor was relatively consistent in 3C-SiC films with thicknesses larger than 280 nm. On the other hand, in the SiC films with thickness below 130 nm, the gauge factor decreased significantly (by 65.9% for 130 nm and and 83.9% for 80 nm film in comparison the 1 μm film).

It should be pointed out that since all as-grown films were thicker than 80 nm, the quantum confinement is ineffective [17]. The crystal defects, on the other hand, should be taken into account due to the following reasons. Owing to the difference

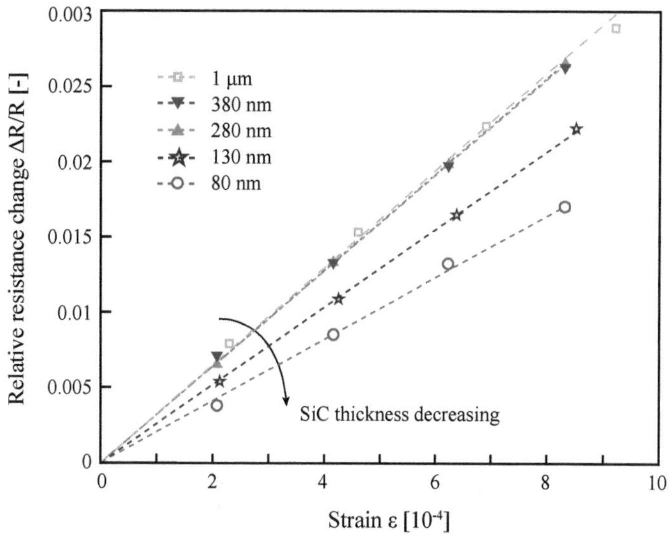

Fig. 4.12 The resistance change of p-type 3C-SiC under strain. Under the same applied strain, 1 μm, 380 and 280 nm films showed almost the same resistance changes ($\Delta R/R$), whereas 130 and 80 nm thin films exhibited considerably smaller resistance change

in lattice size and thermal expansion between SiC and Si, crystal defects typically exist in the epitaxial SiC, which could change the properties of the films. As such, the crystal defects were reported to cause significant impacts on the electrical and mechanical properties of the single crystal 3C-SiC [11, 23]. As presented in chapter [24], the crystalline quality of the 3C-SiC films can be quantified from the FWHM of the rocking curve scan. As such, the reduction of the FWHM in thicker wafers indicated the improvement of crystal quality with increasing film thickness. Furthermore, the mobility factor which was obtained from the hot probe measurement can also be employed to evaluate the influence of crystal defect. Accordingly, 80 nm film exhibited a smaller carrier mobility than that of the 280 nm film ($\sim$7.5cm^2/Vs compared to $\sim$15cm^2/Vs), implying a considerable impact of defects on the properties of the epitaxial 3C-SiC films.

Figure 4.13 shows a cross-sectional TEM of a 3C-SiC film along [110] direction of electron incidence, showing a high density of defect near the SiC/Si interface. These defects were dominantly the stacking faults in [111] orientation, which particularly distributed at the bottom 60 nm from the SiC/Si interface. The TEM image also shows a significant improvement in the film quality with increasing the distance from the SiC/Si interface, which was also observed in other 3C-SiC deposited by different techniques [11, 24]. For the shake of simplicity, the assumptions were made:

(i) As shown in Fig. 4.13, each SiC film includes two layers which are a low density defect layer locating on the top and a high density defect layer underneath. Additionally, because all as-grown 3C-SiC films were deposited using the same

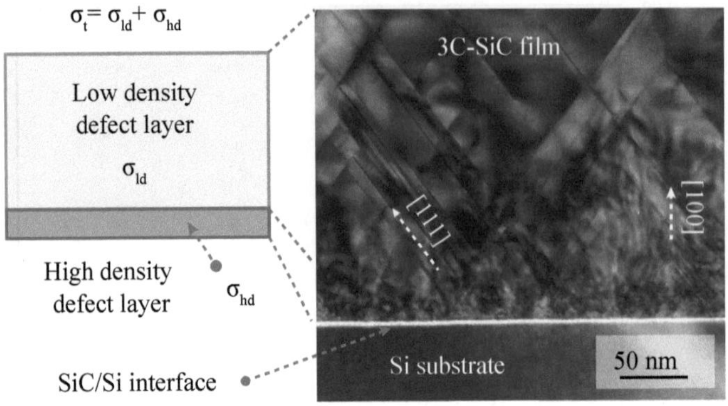

Fig. 4.13 TEM image of a 3C-SiC film

LPCVD parameters, the bottom layer with a high density of defect is expected to exhibit almost the same thickness. Consequently, the film conductance (G_t) is the sum of the conductance of the low density defect layer G_{ld}, and that of the high density defect layer G_{hd}:

$$G_t = G_{ld} + G_{hd} \tag{4.12}$$

(ii) The ratio between the conductance of the high density defect layer and that of the whole SiC film (G_{hd}/G_t) can be approximated as a monotonically decreasing function of the film thickness (t_{SiC}). That means $f(t)$ decreases with increasing t (see Appendix D).

$$G_{hd}/G_t = f(t_{SiC}) \tag{4.13}$$

Under strain, the conductance changes of the high density and the low density defect layers are denoted as ΔG_{hd} and ΔG_{ld}, respectively. As a result, the gauge factor of the SiC is:

$$GF = \frac{\Delta R}{R} \times \frac{1}{\varepsilon} \approx -\frac{\Delta G_t}{G_t} \times \frac{1}{\varepsilon} \tag{4.14}$$

Consequently, the gauge factor can be deduced using the following equation (see Appendix D)

$$\begin{aligned} GF &= -\frac{\Delta G_{ld} + \Delta G_{hd}}{G_t} \times \frac{1}{\varepsilon} = \frac{G_{hd}}{G_t} GF_{hd} + \left(1 - \frac{G_{hd}}{G_t}\right) GF_{ld} \\ &= f(t_{SiC}) GF_{hd} + (1 - f(t_{SiC})) GF_{ld} \end{aligned} \tag{4.15}$$

where GF_{ld} and GF_{hd} are the gauge factor of the low density defect layer and high density defect layer layer, respectively. Evidently, in a thick SiC film, the conductance of high density defect layer is much larger than that of low density defect layer,

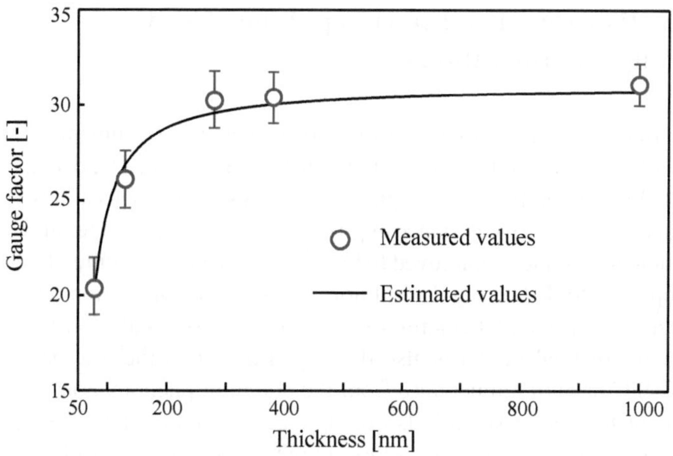

Fig. 4.14 A comparison between estimated values and the experimental results of the gauge factor

resulting in a large G_{hd}/G_t ratio (see Eq. 4.13). Consequently, the gauge factor in those thick films can be considered to be approximately the same as that of the low density defect layer GF_{ld}.

Figure 4.14 shows the estimated gauge factor of p-type 3C-SiC film with different thicknesses, calculated using Eq. 4.15. The experiment data relatively matched with the calculated results employing the proposed model. Therefore, the crystal defect is considered to significantly reduce the magnitude of the piezoresistance in p-type single crystal 3C-SiC film with thickness below 130 nm. For sufficiently thicker films, the impact of defects seems to be less profound. Therefore, the influence of crystal defects should be considered when designing MEMS mechanical sensors employing the piezoresistance of nano thin 3C-SiC films. For instance, to reduce the impact of crystal defects, it is important to minimize the density of defects in the piezoresistive element. Thus, the following method is suggested to enhance the quality of the thin piezoresistive layer. As shown in Fig. 4.13, a high density of defect mainly distributes at the SiC/Si interface, and the concentration of defects significantly decreases with increasing the SiC thickness. Therefore, a high quality nano thin film piezoresistive SiC layer can be obtained by selectively doping the top layers of a thick Si film. The bottom layer of SiC film should be intrinsic. This intrinsic layer can be eliminated using electrochemical mechanical polishing (EMP), leaving the high quality nano thin piezoresistive layer on the top.

4.3 The Piezoresistive Effect in p-Type 3C-SiC at High Temperatures

As SiC based devices are expected to work in harsh environments where the Si based counterparts are not suitable, the behavior of the piezoresistance in p-type 3C-SiC at elevated temperatures requires to be investigated. This aim of this section is to characterize the piezoresistance of p-type 3C-SiC at high temperatures. An novel measurement technique is employed to locally raise the temperature of 3C-SiC resistors by applying Joule heating. In addition, a physical phenomenon on the coupling of the piezoresistance and themoresistive effect in 3C-SiC is also observed. Furthermore, a mathematical model is also developed to obtain the true gauge factor of p-type 3C-SiC at high temperature from the coupling phenomenon. The proposed measurement technique should also be applicable for the characterization of the piezoresistance in other semiconductors at high temperatures as well as be useful in tuning/modifying their gauge factor.

4.3.1 Methodology and Sample Preparation

If the SiC/Si junction is heated, there will be a large leakage current from the SiC layer to the Si substrate, making it impossible to accurately investigate the electrical properties of SiC at high temperatures. The leakage current through this hetero-junction was measured at different temperature varying from 25 °C. up to 200 °C. Experimental data indicated that at low temperature the leakage current through the SiC/Si junction was less than 0.5% of the current flowing through SiC resistors; however at temperatures above 100 °C, this ratio increased above 10%, Fig. 4.15. Consequently, if the whole SiC on Si platform is heated, the Si substrate may has a significantly contribution to the measurement of the piezoresistance of the epitaxial SiC at high temperatures.

Consequently, to avoid the current leakage at the SiC/Si heterojunction for investigating the strain effect in p-type 3C-SiC at high temperatures, the following experimental method was developed, Fig. 4.16. The piezoresistance of p-type SiC was investigated by applying the bending experiment on locally heated SiC resistors using the Joule heating effect. The core idea of this method is supplying a large constant current/voltage to a SiC resistor in order to increase its temperature while maintaining the temperature of the SiC/Si junction as well as the Si substrate near room temperature (approximately 25 °C), at which the leakage current to the substrate is negligible. Subsequently, a mechanical strain is applied once the temperature of SiC resistors has reached a steady state. From the change in SiC resistance under strain at constant current and voltage modes, the magnitude of the piezoresistance in p-type 3C-SiC at high temperatures can be obtained.

Starting from a 150 mm SiC on Si wafer, SiC resistors with dimensions of approximately 12 μm $\times$ 8 μm were fabricated employing a MEMS photolithography process

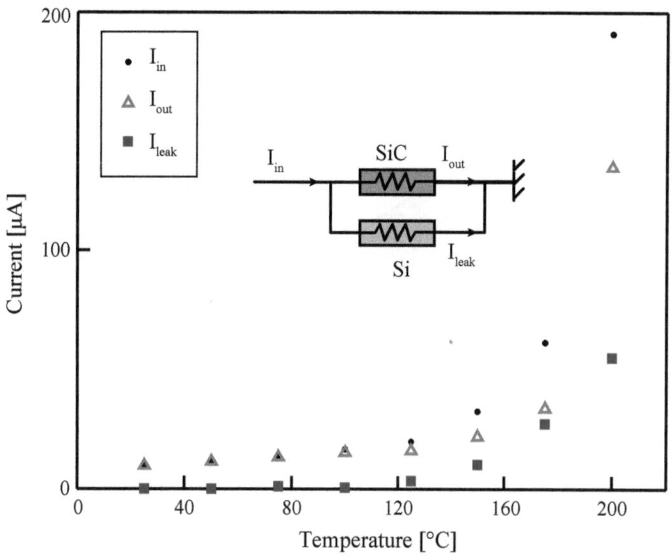

Fig. 4.15 The leakage current through SiC to the Si substrate at high temperatures. In this case, the temperature of the whole SiC on Si sample was increased, leading to a large leakage current through the junction

where aluminum was used to create the electrical contact area. The patterned wafer was then cut into smaller cantilevers with a length of 60 mm and a width of 7 mm for the subsequent bending beam experiment. Finally, to create suspended SiC structure (e.g. SiC bridge), the Si layer underneath was under etched by using XeF_2. Figure 4.17 shows the colorized SEM images of the suspended SiC structures with their dimensions being 200 μm in length and 8μm in width. Due to the anisotropy of the piezoresistance of p-type 3C-SiC [25, 26], all 3C-SiC resistors used in this experiment aligned along the longitudinal [110] direction to obtain a large gauge factor.

4.3.2 The Piezoresistive Effect of the Suspended 3C-SiC at Room Temperature

The piezoresistance of the suspended 3C-SiC resistor was initially investigated at room temperature using the bending beam method. A simulation was also carried out to estimate the strain induced applied to the suspended structures. Accordingly, the strain applied to the suspended structures is comparable to the non-suspended structures. This result is reasonable, since the thickness of SiC is about 0.05% of Si, and the distance from the suspended SiC bridge to the surface of the substrate is much smaller than the that from the surface to the neutral axis of the Si beam.

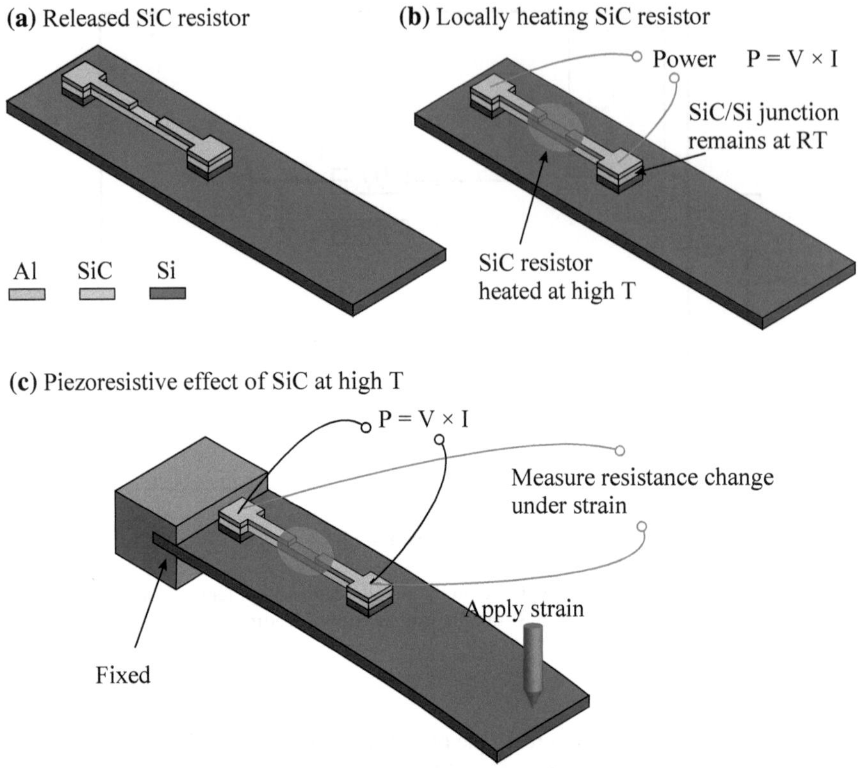

Fig. 4.16 Concept of the proposed methodology for investigating the piezoresistive effect of p-type 3C-SiC at high temperatures. **a** SiC resistors were released from the Si substrate using a photolithography process; **b** A high electrical power was supplied to locally raise the temperature of SiC resistors; **c** Mechanical strain was induced in to SiC resistors, while their resistance change was also measured

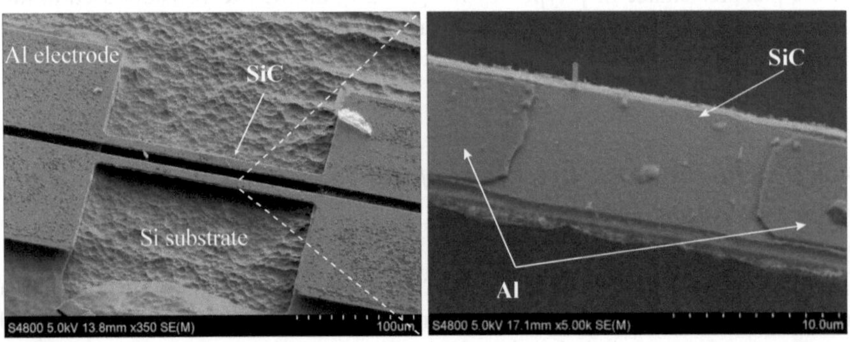

Fig. 4.17 SiC resistors released from Si substrate in which the length of the released bridge is 200 μm, while the length of SiC resistor locating at the center of the bridge is 12 μm

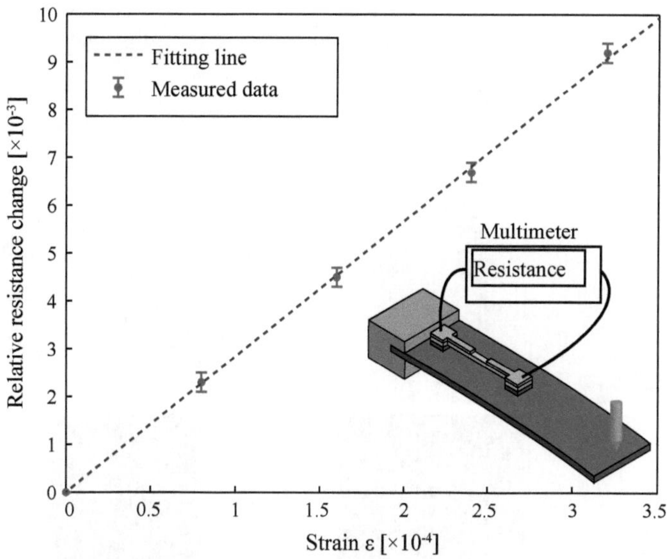

Fig. 4.18 The resistance change of the suspended resistors plotted against applied tensile strain at room temperature. The applied current for resistance measurement was set to be 500 nA, at which Joule heating is expected to be negligible

The resistance of SiC under tensile strain was measured using a resistance-meter (™Agilent Multimeter). A small current of 500 nA was applied to the multimeter to minimize the influence of Joule heating.

Figure 4.18 shows the resistance change of suspended p-type 3C-SiC under tensile strains ranging from 0 to 350 ppm, exhibiting a good linearity. The gauge factor of the suspended p-type 3C-SiC resistors was then found to be 28 at 25 °C, which is comparable to that of the non-released SiC reported in Sect. 4.2.1. This result indicates that the dry etching process of Si employing XeF_2 did not cause any serious damage to the suspended SiC resistors.

4.3.3 The Themoresistive Effect of 3C-SiC

When 3C-SiC resistors are heated, their electrical resistance will change as a result of the change in carrier concentration and charge mobility. This property is named as the themoresistive effect which has been widely adopted in numerous MEMS thermal-based sensors [27, 28]. As such, based on the resistance-temperature (R-T) curve, the temperature of the material or surrounding environments can be estimated by measuring the resistance change. Therefore, the themoresistive effect of p-type 3C-SiC was investigated as a means to monitor the temperature of 3C-SiC resistor when being heated by Joule heating.

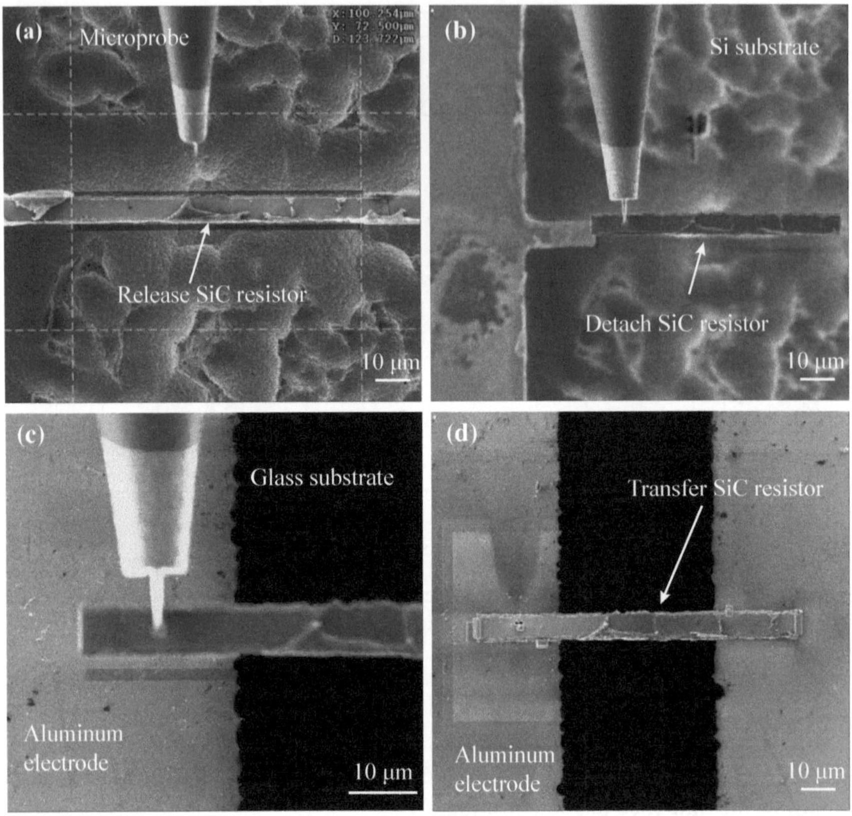

Fig. 4.19 Fabrication process of 3C-SiC resistors transferred onto a glass substrate

To obtain the R-T curve of the p-type 3C-SiC, the influence of the Si substrate at high temperature needs to be eliminated. Consequently, the following fabrication steps were performed to transfer SiC onto an insulator, as illustrated in Fig. 4.19. Firstly, a tungsten probe was attached to the SiC resistors. Secondly, the two sides of the suspended SiC bridges were cut employing FIB to disconnect them from the supporting areas. Next, the SiC bridges were transferred onto an insulator (glass) with aluminum contacts already deposited on it. Finally, to enhance electrical contact, FIB tungsten-deposition was employed to connect the aluminum pads on the glass wafer to the aluminum electrodes on the SiC bridges.

Subsequently, the themoresistive effect of the transferred SiC samples was characterized in a high temperature oven. Figure 4.20 shows the resistance change of SiC on glass against temperatures. Evidently, the p-type 3C-SiC resistances decreased by approximately 50% when the temperature increased from 300 to 600 K This indicated a negative temperature coefficient of resistance (TCR) in the p-type 3C-SiC, which is caused by a increase in carrier concentration due to thermally activation. The R-T curve shown in Fig. 4.20 is then used as a reference for monitoring the temperature of the SiC bridges being heated by Joule effect in the subsequent experiments.

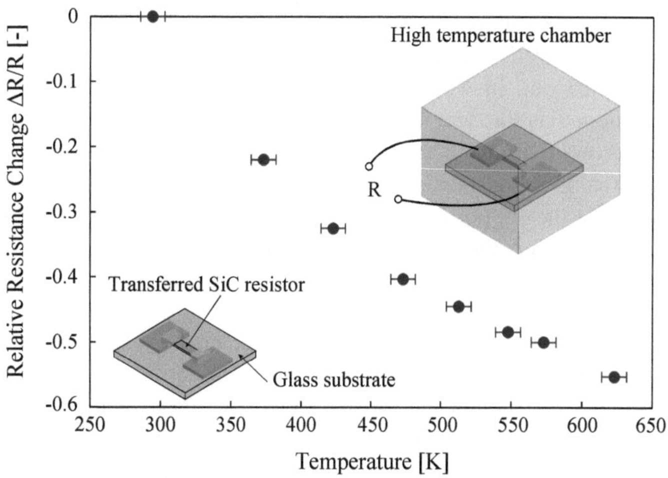

Fig. 4.20 The resistance-temperature (R-T) curve of p-type 3C-SiC

4.3.4 Joule Heating Effect in Suspended SiC Structures

As mentioned above, to diminish the affects of the leakage current from SiC to Si at elevated temperatures, Joule heating was utilized to locally increase the temperature of SiC piezoresistors. The Joule heating effect has also been applied in several MEMS devices including heaters and flow sensors [28–30]. Especially, in micro/nano devices, the capability to raise the temperature at a specific area could lead to low thermal loss, and thus save the energy for the devices.

The physical model of Joule heating in the suspended SiC bridge is developed utilizing the well-established suspended beam heaters for gas sensors [30, 31]. Accordingly, the supplied heating power is $P_{sup} = V \times I$, where V and I are the voltage and current applied to a SiC resistor, respectively. In addition, the supplied heating power will be balanced by the heat losses in the suspended SiC bridge, which include the thermal conduction, thermal convection, and thermal radiation. The heat loss due to thermal conduction along the suspended bridge is given by [30]: $P_{cd} \propto A_{cr}k_{cd}\Delta T$, where A_{cr} is the cross sectional area of the SiC bridges; k_{cd} is the heat conductivity of SiC; and ΔT is the temperature difference between center of the bridge and beam anchor, respectively. The thermal convection causes a heat loss to the ambient, which is approximated as [30]: $P_{cv} \propto A_{sf}k_{cv}\Delta T$, where A_{sf} is the surface area of the suspended bridges, and k_{cv} is the heat transfer coefficient. Additionally, the heat loss through thermal radiation can be estimated as [32]: $P_{rad} \propto A_{sf}\lambda(T_{\infty}^4 - T_0^4)$, where λ is the Stefan–Boltzmann constant, while T_{∞} and T_0 are the temperature of the SiC bridge and the surrounding air, respectively. When the Joule heating effect reaches equilibrium (e.g. the steady state), the following relationship is established:

$$P_{sup} = V_{\infty} \times I_{\infty} = P_{cd} + P_{cv} + P_{rad} \tag{4.16}$$

where, the subscript ∞ stands for the steady state.

Based on the theoretical analysis above, a simulation of Joule heating applied to SiC was carried out, in which air temperature was set at 25 °C, while the Si substrate functioned as the heat sink. The simulation data employing ™COMSOL Multiphysics shows that under Joule heating, high temperatures mainly distributed at the centre area of the suspended Al/SiC bridge while the temperature gradually decreased towards the supporting area along the longitudinal direction of the suspended beam. In addition, the temperature near SiC/Si heterojunction was kept below 50 °C even when the suspended SiC resistor was heated above 300 °C, as shown in Fig. 4.21.

The Joule heating experiment was then conducted on the suspended SiC resistors, utilizing two different approaches which are constant current and constant voltage modes. Both modes exhibited the same results that, when increasing the heating power, a decrease in the SiC resistance was observed. Figure 4.22a plotted the resistance change of SiC agains the heating power measured at the steady state. By correlating the results in Fig. 4.22a and the R-T curve from Fig. 4.20, the relationship between the temperature and the supplied power at the steady state was developed, as presented in Fig. 4.22b. Evidently, increasing the heating power leads to a significant decrease in SiC resistance as well as a profound increase in temperature.

The leakage current from SiC to Si under high heating powers was also observed, indicating that the leakage current was negligible compared the current applied to the SiC resistor (see Appendix E). This result shows solid agreement with the theoretical analysis that thanks to the Joule heating effect, the temperature increased locally at the SiC resistor, and decreased significantly at the SiC/Si supporting areas. Consequently, Joule heating can raise the temperature of the suspended SiC and keep the current leakage at the junction at a acceptably small value. This result allowed the

Fig. 4.21 Simulation of Joule heating applied to a suspended SiC bridge employing COMSOL Multiphysics

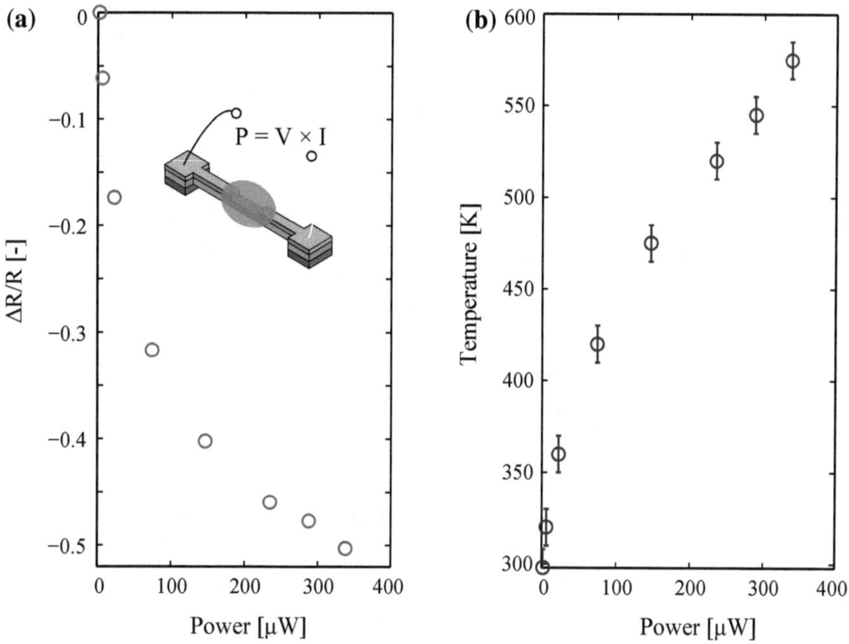

Fig. 4.22 **a** The resistance change of a suspended p-type 3C-SiC resistor against applied powers. **b** The temperature of SiC resistor estimated based on the R-T curve from Fig. 4.20

investigation into the piezoresistance of 3C-SiC on Si substrate at high temperature, which is discussed in the following section.

4.3.5 *Coupling the Piezo- and Thermo-Resistive Effects in SiC*

To characterized the piezoresistance of p-type SiC at elevated temperature, the suspended SiC bridges were initially heated by Joule heating using constant current and constant voltage modes. Once the steady state (T_{∞}) is established, a tensile strain was applied by deflecting the SiC/Si beam downward, and then the resistance changes of the suspended SiC resistor under strain were observed. Subsequently, the gauge factors of the p-type 3C-SiC at temperature T_{∞} were estimated based on the resistance change ($\Delta R/R$).

Experimental data showed that, at a low heating power (below 6 μW) where Joule heating was consider to be insignificant, the change of SiC resistance under tensile strain in the two modes were almost identical. The gauge factor of SiC under this supplied power was found to be about 28, which equal to that measured by the multimeter, as presented in Sect. 4.2.1. Nevertheless, under high heating powers

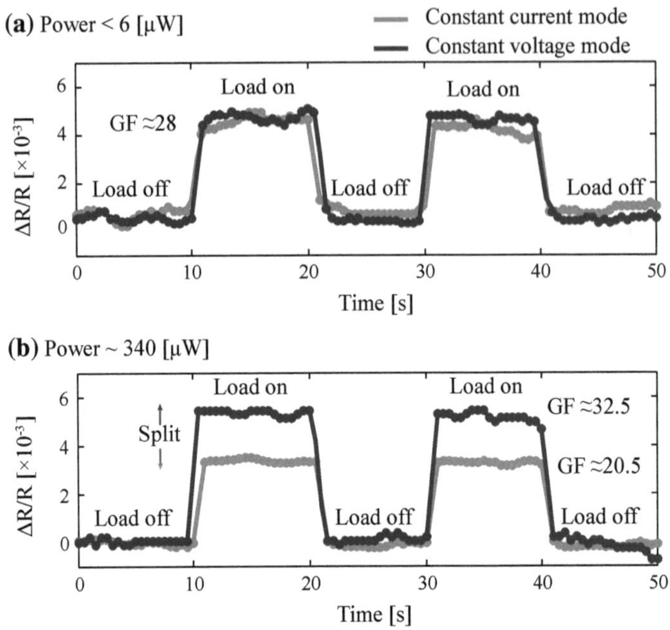

Fig. 4.23 Characterization of the piezoresistance at high temperature employing Joule heating. The resistance change of SiC at **a** low heating power **b** high heating power. The red line illustrates the output of the constant current mode, while the blue line shows the output of the constant voltage mode

(e.g. above 340 μW), the constant current mode exhibited different resistance change with respect to the constant voltage. It is evident from Fig. 4.23 that even under the same mechanical load and heating power, the resistance change of SiC using the constant voltage mode was larger than that measured with the constant current mode. As such, at an heating power of 340 μW corresponding to a temperature of about 573 K according to Fig. 4.22, the constant voltage mode showed a gauge factor of $GF_V = 32.5$ which was considerably larger than that of the constant current mode ($GF_I = 20.5$). This difference in the gauge factor was considered to be reasoned from a coupling effect between the piezoresistance and thermoresistance in the locally heated SiC piezoresistor under external strains.

Figure 4.24a qualitatively presents the coupling phenomenon. Accordingly, under Joule heating, the temperature of SiC initially increased to T at the steady state, and its resistance decreased to R_0 due to the thermoresistance. Subsequently, when temperature remained at T, applying a mechanical strain will cause the SiC resistance to increase R_* following the piezoresistance. In addition, this increase in the resistance will modify the heating power applied to the SiC piezoresistor. For the constant current mode, the heating power will increase as it is proportional to the SiC resistance ($P_I = RI^2$), whereas when applying the constant voltage mode, the heating power will decrease ($P_V = V^2/R$), Fig. 4.24a. The change of the heating power in turn

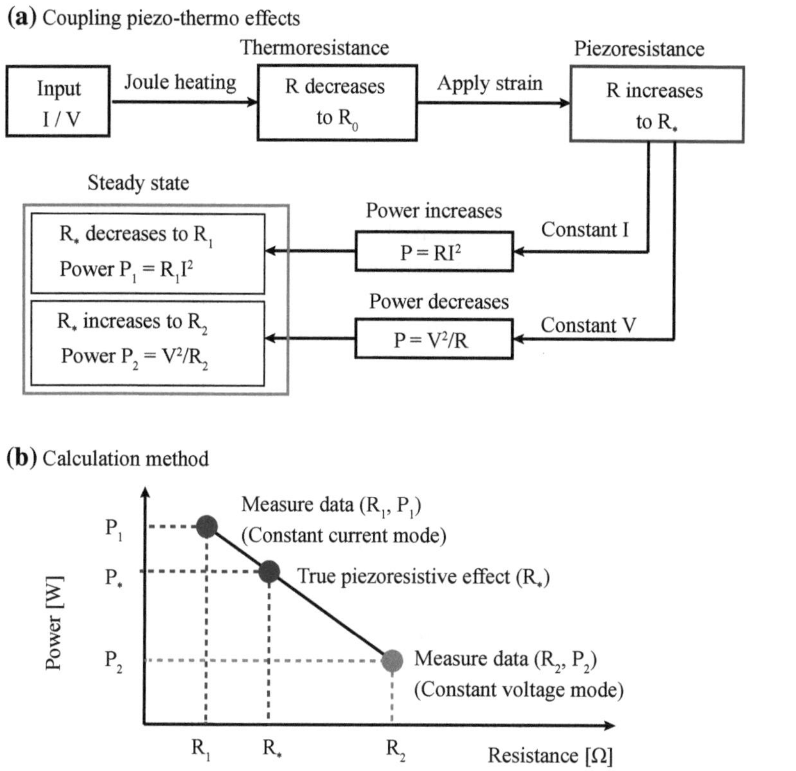

Fig. 4.24 **a** The interaction of the thermoresistance and piezoresistance in a locally heated SiC resistors under strain; **b** The calculation method for the true gauge factor

leads to a change in the SiC resistance due to the thermoresistance. Therefore, at the steady state, the SiC resistance measured at the constant current mode will drop from R_* to R_1; on the other hand, the resistance measured at the voltage mode will increase from R_* to R_2 due to the negative TCR, as presented in Fig. 4.23b.

From the interaction between the piezoresistance and thermoresistance, as well as the measured gauge factors using the current and voltage modes, the true piezoresistive effect at elevated temperature was extracted using the following method. Assuming that under a mechanical strain, if the heating power could be constantly maintained at $P_* = P_0$, the temperature of SiC would be then kept at the same value, and thus the SiC resistance would also remain at R_* at the steady state, Fig. 4.24. Additionally, the true gauge factor of the piezoresistance at T is expected to be sandwiched between GF_I and GF_V because the interaction between the thermoresistive and piezoresistive effects will enhance the gauge factor of SiC at the constant voltage mode; and conversely reduces the gauge factor at the constant current mode. Furthermore, as resistance change of SiC at the steady state was below 0.6%, which resulted in a small change of the heating power of below 0.6%, it is possible to assume a linear relationship between the applied power and the SiC resistance within this small interval (see Appendix E). This linear behavior is schematically illustrated in Fig. 4.24b. Therefore, the equation below was established:

$$\frac{P_1 - P_*}{P_* - P_2} = \frac{R_1 - R_*}{R_* - R_2} \tag{4.17}$$

where P_* is the heating power at the steady state before applying strain ($P_* = P_0 = V^2/R_0 = I^2 R_0$); while P_1 and P_2 are the heating power at the steady state under strain in accordance with the constant current, and constant voltage modes, respectively. Given that δR be the relative resistance change of SiC due to strain, then: $\delta R_1 = (R_1 - R_0)/R_0$, $\delta R_2 = (R_2 - R_0)/R_0$, and $\delta R_* = (R_* - R_0)/R_0$. Equation 4.17 is deduced as as follows:

$$\frac{P_1 - P_*}{P_* - P_2} = \frac{\delta R_1 - \delta R_*}{\delta R_* - \delta R_2} \tag{4.18}$$

In addition, the power differences are given by:

$$\begin{cases} P_1 - P_* = I^2(R_1 - R_0) = \delta R_1 \times P_* \\ P_* - P_2 = V^2(1/R_0 - 1/R_2) \approx \delta R_2 \times P_* \end{cases} \tag{4.19}$$

Consequently, Eqs. 4.18 and 4.19 result in:

$$\frac{\delta R_1}{\delta R_2} = \frac{\delta R_1 - \delta R_*}{\delta R_* - \delta R_2} \tag{4.20}$$

Therefore, the true relative resistance change caused by the piezoresistance of SiC at temperature T can be estimated as:

$$\delta R_* = \frac{2\delta R_1 \delta R_2}{\delta R_1 + \delta R_2} \tag{4.21}$$

Note that as the gauge factor is defined as $GF = \delta R/\varepsilon$, the true gauge factor of the piezoresistance at temperature T can be obtained from the measured gauge factors at the constant current and voltage modes:

$$GF_* = \frac{\delta R_*}{\varepsilon} = \frac{2(\delta R_1/\varepsilon) \times (\delta R_2/\varepsilon)}{\delta R_1/\varepsilon + \delta R_2/\varepsilon} = \frac{2GF_\mathrm{I} GF_\mathrm{V}}{GF_\mathrm{I} + GF_\mathrm{V}} \tag{4.22}$$

Figure 4.25 plots the gauge factors of p-type single crystal 3C-SiC at different temperature calculated using Employing Eq. 4.22. It is evident that, the gauge factor was relatively stable with with temperature ranging from 298 to 573 K, exhibiting a small deviation of below 10%. This result is consideration, since the p-type 3C-SiC was relatively highly doped (5×10^{18} cm^{-3}), which makes its piezoresistance more stable at high temperatures. Additionally, the results indicate the feasibility of using 3C-SiC for sensing applications at temperatures well-above the typical operating temperatures of Si. Moreover, it was also clear that applying a sufficiently high voltage mode could lead to an increase in the gauge factor of SiC by approximately 20%

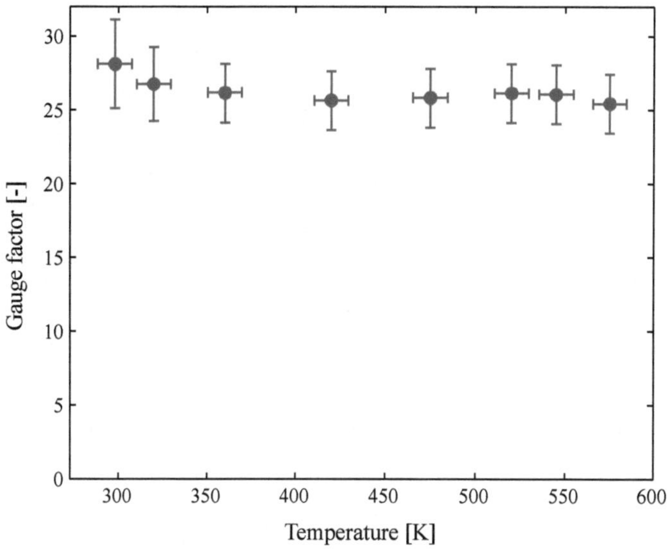

Fig. 4.25 Calculated gauge factor of SiC at different heating temperatures

with respect to the true gauge factor. It means that the coupling of the piezoresistance and thermoresistance is applicable for tuning or enhancing the sensitivity of SiC transducers.

4.4 The Piezoresistive Effect in Four-Terminal SiC Resistors

As described in Chap. 2, there are two types of piezoresistors, which are the two terminal, and four-terminal resistors [33, 34]. The piezoresistance of four-terminal resistors is also named as the pseudo-Hall effect, as applying a shear strain could generate a voltage across two terminals of a four-terminal device due to the distortion of potential distribution [35–37]. There have been numerous studies on the piezoresistance of four-terminal Si devices, and in fact this effect has been applied in commercial products such as strain sensors [37, 38]. This section presents the piezoresistance of four-terminal p-type 3C-SiC resistors as well as its orientation dependence. The shear piezoresistive coefficients, π_{61}, π_{62}, and π_{66}, which define the magnitude of the effect were also characterized.

4.4.1 Configuration of SiC Four-Terminal Resistors

Figure 4.26 shows SiC four-terminal resistors used in this study, fabricated using the similar process as the two terminal resistors reported in Chap. 3. All four-terminal

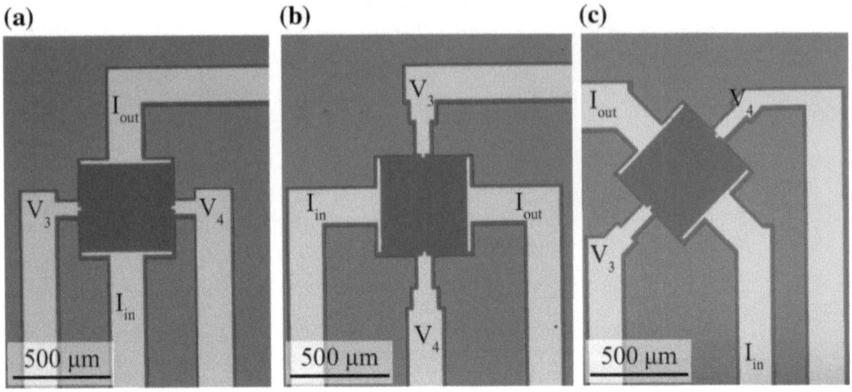

Fig. 4.26 SiC four-terminal resistors with dimensions of 500 μm × 500 μm × 280 nm, aligned in different orientations

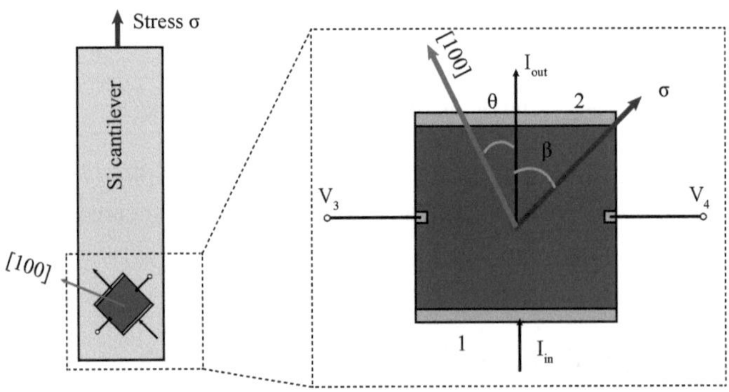

Fig. 4.27 The configuration of four-terminal resistors including the directions of the applied current, measured voltage and applied stress

resistors were formed in a square shape with dimensions of 500 μm in length, 500 μm in width and 280 nm in thickness. In addition, in each SiC four-terminal resistors, the two line-electrodes were employed to supply current, whereas the other two point-electrodes were utilized to measure the output voltage. Figure 4.27 illustrates the configuration of a four-terminal resistors, in which θ is the angle between the direction of the applied current and [100] orientation, whereas β is the angle between the directions of the applied stress and the applied current. Furthermore, to characterize the shear piezoresistive coefficients as well as the orientation dependence, SiC four-terminal resistors were fabricated in different orientations, and then stress was also induced in different directions, as listed in Table 4.4.

A constant current was applied through terminals 1 and 2, and the generated voltage across terminal 3 and 4 under stress was monitored. The strain induced into the SiC four-terminal devices was calculated the method presented in Sect. 4.1.

Table 4.4 List of SiC four-terminal samples

Test samples	Orientation of current	Orientation of uniaxial stress	θ	β
A	[100]	[110]	0°	45°
B	[110]	[110]	45°	0°
C	[110]	$[1\bar{1}0]$	45°	90°
D	[110]	[100]	45°	45°
E	[100]	[100]	0°	0°
F	[100]	[010]	0°	90°

4.4.2 *Shear Piezoresistive Coefficients in Four-Terminal Resistors*

The experiment was firstly conducted on sample A, in which the orientations of the current was [100], and that of the mechanical stress was [110], respectively. A current of 10 μA was applied using a ™Agilent U2722A current source, while the generated voltage was measured using a ™Agilent MSO-X 3104A oscilloscope. Figure 4.28 shows that a voltage was generated at terminal 3, 4 of sample A when a tensile stress is applied, and this generated voltage increased when the applied stress increased.

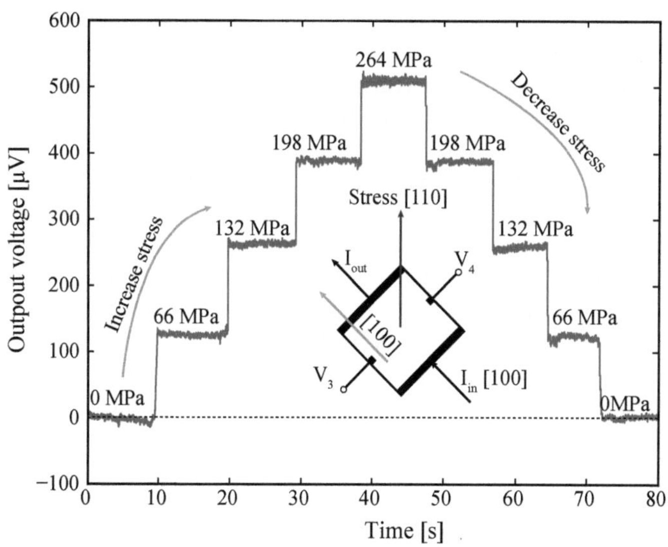

Fig. 4.28 The generated voltage measured at terminals 3 and 4 while a constant current of 10 μA was applied through terminals 1 and 2 under stresses ranging from 0 to 264 MPa. The orientation of the current of the device was [100], while that of the applied stress was [110]

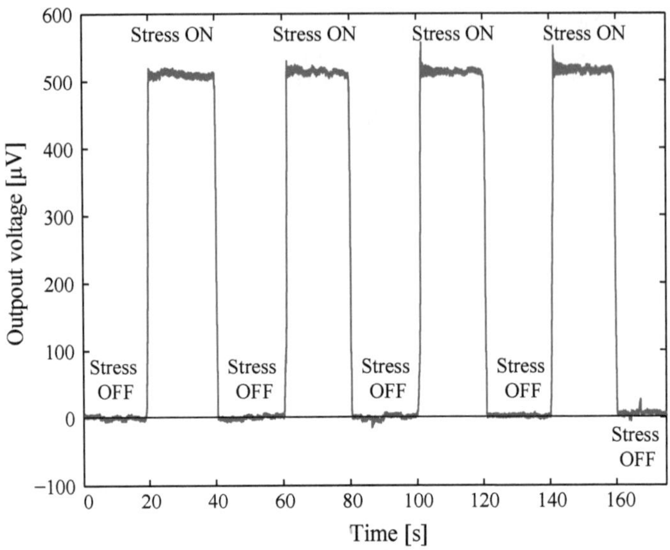

Fig. 4.29 The generated voltage across terminals 3 and 4 under several loading cycles at a constant current of 10 μA, and a constant stress of 264 MPa

The generated voltage then decreased when the stress decreased, and returned to 0 after completely removing the applied stress.

The generated voltage of the SiC four-terminal resistors was observed for several stress-applying cycles, as shown in Fig. 4.29. Under a tensile stress of 240 MPa, the generated voltage of sample A was found to be 510 μV, exhibiting a good reproducibility without any significant drift voltage.

The same experiments were also performed on other samples with θ and β described in Table 4.4. The same phenomenon was also observed in other samples (B, C, D, E, F) that a voltage was generated across terminals 3 and 4 under a mechanical stress. Figure 4.30 plots the ratio of the output voltage to the input voltages of different samples A, B, C, D, E and F (V_{out}/V_{in}) against the applied stress (σ). In addition, it is evident that (V_{out}/V_{in} varies with the directions of the current (θ) and the applied stress (β), indicating an anisotropic property. The model developed by Kanda et al. for Si [40, 41] was employed to analyze the orientation dependence of the piezoresistance in SiC four-terminal resistors, due to the similar in crystal structure of the two materials. Figure 4.30 (Inset) shows an equivalent circuit of a four-terminal device, consisting of four resistors R_{13}, R_{14}, R_{23}, and R_{24}. When no strain is applied, these four resistors are considered to be identical ($R_{13} = R_{14} = R_{23} = R_{24}$), resulting in a 0 output voltage. When applying a uniaxial strain as shown in the inset of Fig. 4.30, strains are induced into these four resistors, changing their resistance. As R_{13}, R_{14}, R_{23}, and R_{24} are aligned in different orientations, these resistance changes are different, breaking the balance of the bridge circuit, which causes a non-zero voltage at terminals 3 and 4. The ratio of the generated voltage (V_{out}) across terminals 3 and

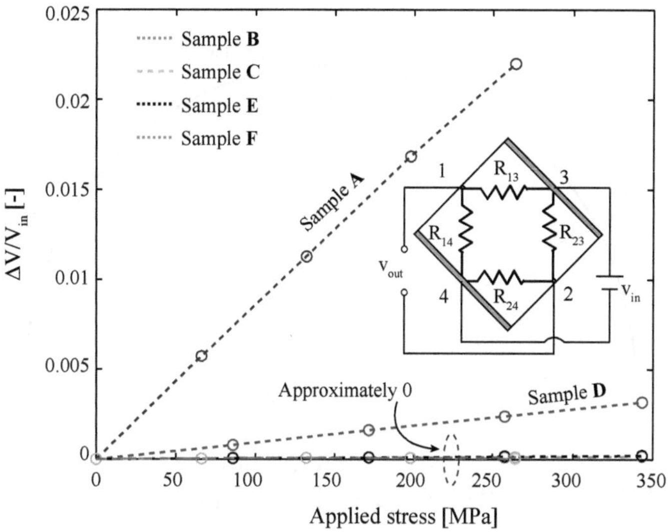

Fig. 4.30 The ratio of the generated voltage and applied voltage of the SiC four-terminal devices aligned in different orientation. *Inset* The equivalent four resistors R_{13}, R_{14}, R_{23}, and R_{24} of a 3C-SiC four-terminal device [39]

4 to the applied voltage across terminals 1 and 2 (V_{in}) under stress can be quantified by the coefficients π_{61}, π_{62}, and π_{66}, as expressed follows [39, 42]:

$$V_{out} = V_{in}(\pi_{61}\sigma_1 + \pi_{62}\sigma_2 + \pi_{66}\sigma_6) \tag{4.23}$$

where σ_1 and σ_2 are normal stresses which are parallel and perpendicular to the current; and σ_6 is the in-plane shear stress which can be obtained from the applied uniaxial stress σ, using Mohr's circle (see Appendix F):

$$\begin{cases} \sigma_1 = \sigma \cos^2 \beta \\ \sigma_2 = \sigma \sin^2 \beta \\ \sigma_6 = -\dfrac{\sin 2\beta}{2}\sigma \end{cases} \tag{4.24}$$

The shear piezoresistive coefficients are defined in the following equation [41, 43] (see Appendix F):

$$\begin{cases} \pi_{61} = -\dfrac{1}{4}a \sin 4\theta \\ \pi_{62} = \dfrac{1}{4}a \sin 4\theta \\ \pi_{66} = b + a \sin^2 2\theta \end{cases} \tag{4.25}$$

where a and b are constants which can be obtained from the change of the generated voltage in samples A and D. Consequently, the shear piezoresistive coefficients

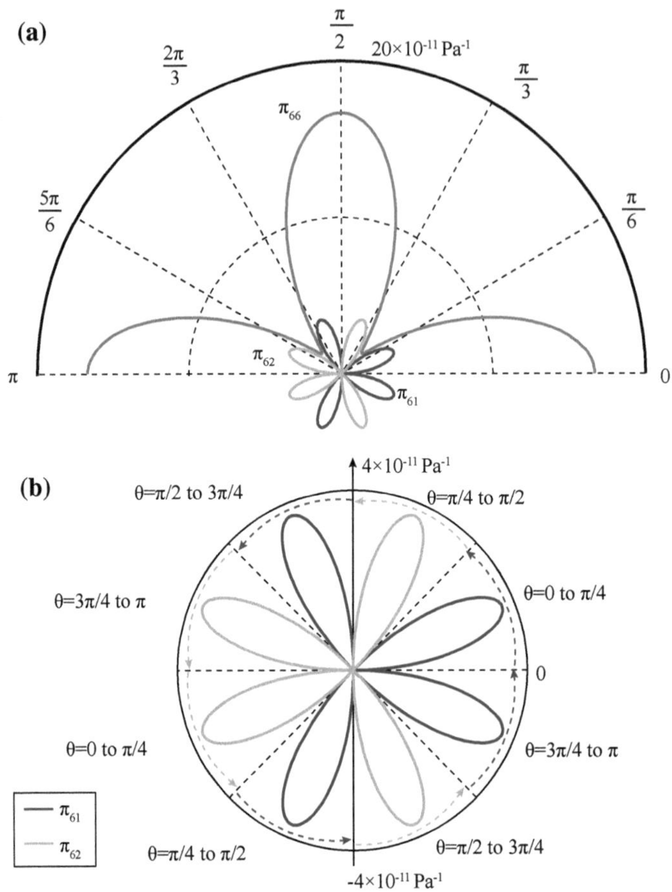

Fig. 4.31 **a** The coefficients π_{61}, π_{62}, and π_{66} of SiC four-terminal devices when θ varies from 0 to π; **b** π_{61} and π_{62} are plotted in a different scale

π_{61}, π_{62} and π_{66} can also be deduced. These piezoresistive coefficients can also be determined from the fundamental piezoresistive coefficients π_{11}, π_{12}, and π_{44}, as presented in Chap. 2. From the experimental results shown in Fig. 4.30, a and b were calculated to be -14.8×10^{-11} Pa^{-1} and 16.7×10^{-11} Pa^{-1}, respectively. Furthermore, substituting a and b into Eq. 4.25, the magnitude of the coefficients in any arbitrary orientation on (100) plane can be estimated, as shown in Fig. 4.31. These results indicate that for the p-type 3C-SiC four-terminal devices, the coefficient π_{66} plays a more dominant role than π_{61} and π_{62}. The experimental results of samples B, C, E, F were also in good agreement with the theoretical calculation that the generated voltage was approximately 0, since the coefficients π_{61} and π_{62} are 0 when θ equals 0 or $\pi/4$ radian.

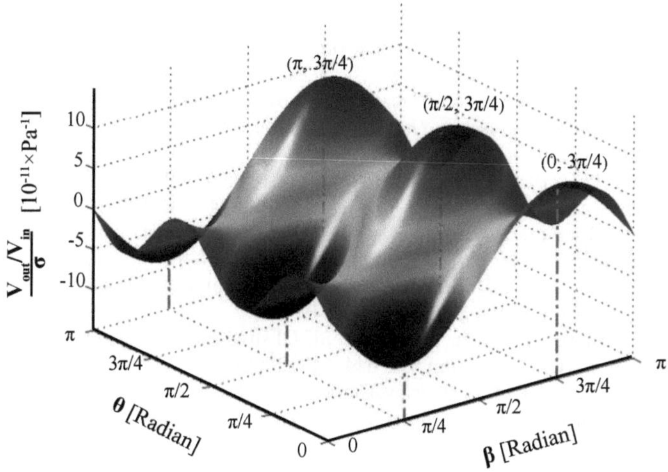

Fig. 4.32 The ratio of the generated voltage across terminals 3 and 4 to the input voltage across terminals 1 and 2 per unit uniaxial stress σ

4.4.3 Orientation Dependence of the Piezoresistive Effect in Four-Terminal Resistors

Based on these coefficients, the magnitude of the output offset voltage of 3C-SiC four-terminal devices under stress was estimated. Substituting Eq. 4.24 into 4.23, and using the calculated piezoresistive coefficients shown in Fig. 4.31, the magnitude of the piezoresistance of 3C-SiC four-terminal resistors aligned in arbitrary orientations were theoretically estimated. Figure 4.32 plots $([V_{out}/V_{in}]/\sigma)$ against applied unit strain. Evidently, the magnitude of the effect in p-type 3C-SiC four-terminal resistors depends on the orientations of both current and stress. As such, in (100) plane, the generated voltage ($|V_{out}|$) is maximized at $(\theta,\ \beta) = (m\pi/2,\ (2n+1)\pi/4)$, where m and n are integers. On the other hand, for the Hall-devices where a constant offset voltage under external stress is desired, the orientations of $(\theta,\ \beta) = (m\pi/4,\ n\pi/2)$ should be selected to minimize the piezoresistance of four-terminal resistors (here m and n are integers).

References

1. R.G. Ballas, *Piezoelectric Multilayer Beam Bending Actuators: Static and Dynamic Behavior and Aspects of Sensor Integration* (Springer Science & Business Media, Berlin, 2007)
2. X. Li, W.Y. Shih, I.A. Aksay, W.E. Shih, Electromechanical behavior of PZT-brass unimorphs. J. Am. Ceram. Soc. **82**(7), 1733 (1999)
3. X. Gao, W.E. Shih, W.Y. Shih, Induced voltage of piezoelectric unimorph cantilevers of different nonpiezoelectric/piezoelectric length ratios. Smart Mater. Struct. **18**, 125018 (2009)

4. F. Iacopi, G. Walker, L. Wang, L. Malesys, S. Ma, B.V. Cunning, A. Iacopi, Orientation-dependent stress relaxation in hetero-epitaxial 3C-SiC films. Appl. Phys. Lett. **102**, 011908 (2013)
5. M.D. Nguyen, H.-P. Phan, K. Matsumoto, I. Shimoyama, A sensitive liquid-cantilever diaphragm for pressure sensor, in *The 26th IEEE International Conference on MEMS* (Taipei, Taiwan, 2013), pp. 617–620
6. M. Tortonese, R.C. Barrett, C.F. Quate, Atomic resolution with an atomic force microscope using piezoresistive detection. Appl. Phys. Lett. **62**(8), 834–836 (1993)
7. T. Homma, K. Kamimura, H.Y. Cai et al., Preparation of polycrystalline SiC films for sensors used at high temperature. Sens. Actuators. A **40**(2), 93–96 (1994)
8. J.S. Shor, D. Goldstein, A.D. Kurtz, Characterization of n-type β-Sic as a piezoresistor. IEEE Trans. Electron Devices **40**(6), 1093–1099 (1993)
9. K. Yasui, H. Miura, M. Takata, T. Akadane, SiCOI structure fabricated by catalytic chemical vapor deposition. Thin Solid Film **516**, 644–647 (2008)
10. C.H. Wu, C.A. Zorman, M. Mehregany, Fabrication and testing of bulk micromachined silicon carbide piezoresistive pressure sensors for high temperature applications. IEEE Sens. J. **6**(2), 316–324 (2006)
11. M. Eickhoff, M. Stutzmann, Influence of crystal defects on the piezoresistive properties of 3C-SiC. J. Appl. Phys. **96**, 2878 (2004)
12. T. Akiyama, D. Briand, N.F. de Rooiji, Design-dependent gauge factors of highly doped n-type 4H-SiC piezoresistors. J. Micromech. Microeng. **22**, 085034 (2012)
13. J.S. Shor, L. Bemis, A.D. Kurtz, Characterization of monolithic n-type 6H-SiC piezoresistive sensing elements. IEEE Trans. Electron Devices **41**(5), 661–665 (1994)
14. R.S. Okojie, A.A. Ned, A.D. Kurtz et al., Characterization of highly doped n and p-type 6H-SiC piezoresistors. IEEE Trans. Electron Devices **45**(4), 785–790 (1998)
15. Y. Kanda, Piezoresistance effect of silicon. Sens. Actuators A **28**, 83 (1991)
16. Y. Sun, S.E. Thompson, T. Nishida, *Strain Effect in Semiconductor: Theory and Device Applications*, 1st edn. (Springer, Berlin, 2009)
17. R. He, P. Yang, Giant the piezoresistive effect effect in silicon nanowires. Nat. Nanotech. **1**, 42–46 (2006)
18. Y. Yang, X. Li, Giant piezoresistance of p-type nano-thick silicon induced by interface electron trapping instead of 2D quantum confinement. Nanotechnology **22**, 015501 (2011)
19. J. Bi, G. Wei, L. Wang, F. Gao, J. Zheng, B. Tang, W. Yang, Highly sensitive piezoresistance behaviors of n-type 3C-SiC nanowires. J. Mater. Chem. C **1**, 4514 (2013)
20. F. Gao, J. Zheng, M. Wang, G. Wei, W. Yang, Piezoresistance behaviors of p-type 6H-SiC nanowires. Chem. Commun. **47**, 11993–11995 (2011)
21. R. Shao, K. Zheng, Y. Zhang, Y. Li, Z. Zhang, The piezoresistive effect behaviors of ultra-strained SiC nanowires. Appl. Phys. Lett. **101**, 233109 (2012)
22. P. Tanner, L. Wang, S. Dimitrijev, J. Han, A. Iacopi, L. Hold, G. Walker, Novel electrical characterization of thin 3C-SiC films on Si substrates. Sci. Adv. Mater. **6**(7), 1542–1547 (2014)
23. X. Song, J.F. Michaud, F. Cayrel, M. Zielinski, M. Portail, T. Chassagne, E. Collard, D. Alquier, Evidence of electrical activity of extended defects in 3C-SiC grown on Si. Appl. Phys. Lett. **96**, 142104 (2010)
24. L. Wang, S. Dimitrijev, J. Han et al., Demonstration of p-type 3C-SiC grown on 150 mm Si(100) substrates by atomic- layer epitaxy at 1000 °C. J. Cryst. Growth **329**, 67–70 (2011)
25. H.-P. Phan, D.V. Dao, K. Nakamura, S. Dimitrijev, N.-T. Nguyen, The piezoresistive effect of SiC for MEMS sensors at high temperatures: a review. J. Microelectromech. Syst. **24**(6), 1663–1677 (2015)
26. H.-P. Phan, A. Qamar, D.V. Dao, T. Dinh, L. Wang, J. Han, P. Tanner, S. Dimitrijev, N.-T. Nguyen, Orientation dependence of the pseudo-Hall effect in p-type 3C-SiC four-terminal devices under mechanical stress. RSC Adv. **5**(69), 56377–56381 (2015)
27. T. Dinh, D.V. Dao, H.P. Phan, L. Wang, A. Qamar, N.T. Nguyen, P. Tanner, M. Rybachuk, Charge transport and activation energy of amorphous silicon carbide thin film on quartz at elevated temperature. Appl. Phys. Express **8**(6), 061303 (2015)

28. N.T. Nguyen, Micromachined flow sensors-a review. Flow Meas. Instrum. **8**(1), 7–16 (1997)
29. D.V. Dao, V.T. Dau, T. Shiozawa, S. Sugiyama, Development of a dual-axis convective gyroscope with low thermal-induced stress sensing element. J. Microelectromech. Syst. **16**(4), 950–958 (2007)
30. Q. Zhou, A. Sussman, J. Chang, J. Dong, A. Zettl, W. Mickelson, Fast response integrated MEMS microheaters for ultra low power gas detection. Sens. Actuators. A **223**, 67–75 (2015)
31. A. Baranov, D. Spirjakin, S. Akbari, A. Somov, Optimization of power consumption for gas sensor nodes: A survey. Sens. Actuators A **233**, 279–289 (2015)
32. H. Chiriac, I. Astefanoaei, A model of the DC Joule heating in amorphous wires. Physica Status Solidi (a) **153**(1), 183–189 (1996)
33. R.C. Jaeger, J.C. Suhling, R. Ramani, Errors associated with the design, calibration and application of piezoresistive stress sensors in (100) silicon. IEEE Trans. Compon. Hybrids Manufact. Technol. **17**(1), 97–107 (1994)
34. A. Mian, J.C. Suhling, R.C. Jaeger, The van der pauw stress sensor. IEEE Sens. J. **6**(2), 340–356 (2006)
35. S. Yue, W.A. Moussa, L.R. Williston, Feasibility on a piezoresistive tactile normal force sensor array. Appl. Phys. A **116**(2), 409–414 (2014)
36. M. Doelle, D. Mager, P. Ruther, O. Paul, Geometry optimization for planar piezoresistive stress sensors based on the pseudo-Hall effect. Sens. Actuators. A **127**(2), 261–269 (2006)
37. J.C. Doll, B.L. Pruitt, *Piezoresistor Design and Applications* (Springer, Berlin, 2013). ISBN 978-1-4614-8516-2
38. J.J.E. Gragg, Silicon pressure sensor, U.S. Patent 4,317,126, 23, 1982
39. Y. Kanda, K. Yamamura, Four-terminal-gauge quasi-circular and square diaphragm silicon pressure sensors. Sens. Actuators **18**(3), 247–257 (1989)
40. Y. Kanda, Graphical representation of the piezoresistance coefficients in silicon-shear coefficients in plane. Jpn. J. Appl. Phys. **26**(7R), 1031 (1987)
41. Y. Kanda, A graphical representation of the piezoresistance coefficients in silicon. IEEE Trans. Electron Devices **29**(1), 64–70 (1982)
42. D.V. Dao, T. Toriyama, J. Wells, S. Sugiyama, Silicon piezoresistive six-degree of freedom micro force-moment sensor. Sens. Mater. **15**, 113–135 (2002)
43. W.G. Pfann, R.N. Thurston, Semiconducting stress transducers utilizing the transverse and shear piezoresistance effects. J. Appl. Phys. **32**(10), 2008–2019 (1961)

Chapter 5
The Piezoresistive Effect in p-Type Nanocrystalline SiC

Different from single crystalline SiC, nanocrystalline SiC (nc-SiC), with its grain size in sub-micron scale, can be grown on various substrates (e.g. silicon, silicon dioxide, silicon nitride) and therefore, it is a good candidate for MEMS transducers [1–3]. Nanocrystalline SiC can be grown by chemical vapor deposition (CVD) at 1100 °C [3], sputtering amorphous SiC at approximately 750 °C followed by thermal annealing [4], and *in situ* crystallization of amorphous SiC at 600 °C [5]. Recent studies also demonstrated the possibility of synthesis nc-SiC at room temperature using the supersonic molecular beam epitaxy [7]. The capability of growth at low temperatures on various substrates not only simplifies the fabrication, but also allows MEMS surface processing in nc-SiC. However, to date, no study has reported the piezoresistive effect of p-type nc-SiC for MEMS mechanical sensing applications. This section, therefore, aimed at characterizing the piezoresistive effect of p-type nc-SiC and comparing with single crystalline 3C-SiC reported in Chap. 4.

5.1 Growth Process of Nanocrystalline SiC on Si

The nc-SiC films were deposited on n-type (100) Si wafers using the low pressure chemical vapor deposition (LPCVD) method at a low temperature of 600 °C [5]. Prior to deposition, standard RCA cleaning was carried out to clean the surface of the Si wafers. Methylsilane (H_3SiCH_3) was used as a single precursor in a hot wall LPCVD reactor. Trimethylaluminium (TMA, $Al(CH_3)_3$) with a flow rate of 0.15 sccm was introduced as a means of doping p-type SiC. Aluminum (Al) was utilized as the acceptor atom in this *in situ* doping process due to its shallow ionization energy of approximately 200 meV. X-ray Photoelectron Spectroscopy (XPS) analysis showed that the concentration of Al was 4.1 at. %, while the Si/C ratio was 0.68.

The microstructures of deposited SiC films were characterized by transmission electron microscopy (TEM) and x-ray diffractometry (XRD). The thickness of the nc-SiC film was 240 nm and their surface roughness was approximately 6 nm. Figure 5.1 shows a typical SAED pattern and High Resolution TEM (HRTEM)

H.-P. Phan, *Piezoresistive Effect of p-Type Single Crystalline 3C-SiC*,
Springer Theses, DOI 10.1007/978-3-319-55544-7_5

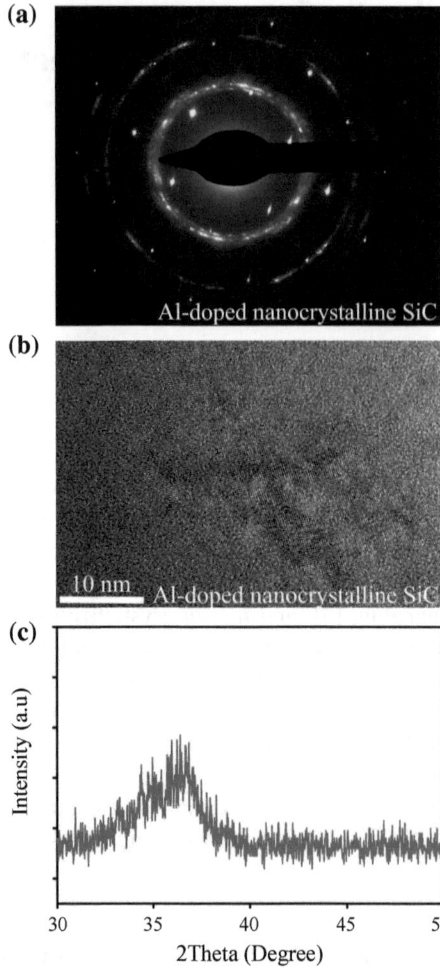

Fig. 5.1 Optical characterization of nc-SiC: **a** SAED pattern—the spotty rings indicate that the film contains randomly oriented nc-SiC grains; the arrayed diffraction spots are from underlying Si substrate; **b** Cross-sectional HRTEM image—the presence of randomly oriented lattice grains and fringes also confirm that the film is made of nc-SiC; **c** X-Ray diffraction pattern—the broad diffraction peak indicates that the SiC grains are relatively small (average size around 13.4 nm) and exhibit some degree of misorientation. The residual stress in the film can also cause the broadening of the diffraction peak. (Figures (a) and (b): Reproduced with permission from [5]. Copyright [2009], AIP Publishing LLC.)

image. Multi-rings with spots are observed in the SAED pattern, whereas the arrayed diffraction spots are from the Si substrate beneath, Fig. 5.1a. These spotty rings and lattice fringes shown in Fig. 5.1b imply that the deposited film is composed of SiC nano grains with different orientations [5]. The broad diffraction peak at around 36.2° from $\theta - 2\theta$ XRD scan indicates that the SiC grains are relatively small and exhibit some degree of misorientation, Fig. 5.1c. This result also suggests that the film might include 3C-SiC(111), 4H-SiC(004), 6H-SiC(102), and other SiC poly types, which can contribute to a diffraction peak at this 2θ position. Additionally, the residual stress in the film can also cause the broadening of the diffraction peak. According to Debye-Scherrer equation [8], the average SiC grain size was found to be around 13.4 nm. The capacitance-voltage (C-V) measurements demonstrated that the SiC is p-type conductive with a doping concentration of $2 \times 10^{18}\,\mathrm{cm}^{-3}$ [5].

5.2 Characterization of the Gauge Factor in p-Type Nanocrystalline SiC

To investigate the piezoresistive effect, nc-SiC piezoresistors on a Si substrate were patterned using a conventional photolithography process, as presented in Chap. 3. The size of the SiC resistors was 100 μm × 300 μm × 240 nm. Aluminum was used as the electrodes of the resistors. The current-voltage (I-V) curve of the SiC resistors shows a linear relationship between the applied voltage and measured current. Accordingly, the resistivity of the nc-SiC was found to be approximately 72 Ωcm^{-1} which is 3 orders of magnitude larger than that of a single crystalline 3C-SiC film at the same carrier concentration [9, 10]. This indicates that the grain boundary of nc-SiC significantly increases the electrical resistance of nc-SiC films. As the p-type nc-SiC was grown on n-type Si, it is important to investigate the contribution of Si substrate to the measurement of the piezoresistive effect in nc-SiC. The pn junction between p-type nc-SiC and n-type Si creates a depletion layer at SiC/Si interface, preventing the leakage-current from SiC to Si layer [11, 12]. The leakage-current through SiC/Si heterojunction was measured to be 2% of the current flowing in SiC resistors. This indicates that, the substrate contributed 2% of the measured conductance of nc-SiC, which needs to be considered when calculating the gauge factor of nc-SiC. To lower the influence of substrate, the n-type Si substrate was aligned in [110] orientation, which has a smaller piezoresistive effect than [100] orientation [13–15] (Fig. 5.2).

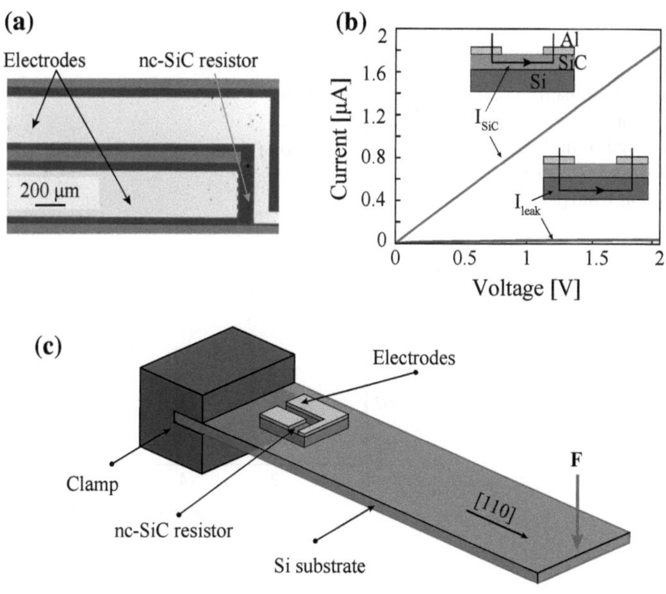

Fig. 5.2 **a** Photograph of a nc-SiC resistor. **b** The I-V curve of nc-SiC resistor. **c** Schematic sketch of the bending experiment. Reprint with permission from [6]

The Si wafer with nc-SiC resistors on it was then diced into strips with a dimension of 60 mm × 9 mm× 0.6 mm and the bending beam method was then employed to induce strain on nc-SiC film. The strain into the nc-SiC was estimated to be in a range of from 0 to 800 ppm when varying the applied force from 0 to 2 N.

The resistance change of the SiC resistor during the bending experiment was then measured by the 4 points measurement using the Agilent™34410A Multimeter. To reduce the influence of the Joule's heating effect, the resistance of nc-SiC was measured at a low current of 500 nA and a consumption power of below 0.25 μW. Figure 5.3 shows the relationship between the applied strain and the relative resistance change of p-type nc-SiC, indicating that the resistance of nc-SiC increased proportionally with increasing strain. The gauge factor of nc-SiC is calculated as:

$$GF = \frac{\Delta R_{\mathrm{SiC}}/R_{\mathrm{SiC}}}{\varepsilon} = -\frac{\Delta\sigma_{\mathrm{SiC}}/\sigma_{\mathrm{SiC}}}{\varepsilon} \tag{5.1}$$

where R_{SiC} and σ_{SiC} are the resistance and conductance of nc-SiC, respectively. Since there is a current leakage from nc-SiC layer to the Si substrate (approximately 2% of the current flowing through the SiC resistor), the influence of the piezoresistive effect of Si substrate and SiC/Si heterojunction is necessarily taken into account. Let σ_{sub} be the conductance of the substrate. The measured conductance (or resistance) is the sum of SiC conductance and substrate conductance:

$$\sigma_{\mathrm{c}} = \sigma_{\mathrm{SiC}} + \sigma_{\mathrm{sub}} \tag{5.2}$$

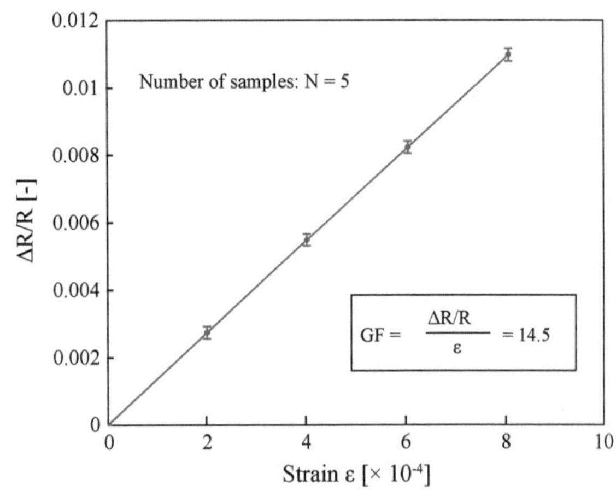

Fig. 5.3 The relationship between the relative resistance change of nc-SiC and applied strains. Reprint with permission from [6]

Therefore, the relative conductance change is:

$$\frac{\Delta\sigma_c}{\sigma_c} = \frac{\Delta\sigma_{SiC} + \Delta\sigma_{sub}}{\sigma_c} = \frac{\sigma_{SiC}}{\sigma_c}\frac{\Delta\sigma_{SiC}}{\sigma_{SiC}} + \frac{\sigma_{sub}}{\sigma_c}\frac{\Delta\sigma_{sub}}{\sigma_{sub}}$$
$$= \frac{98}{100}\frac{\Delta\sigma_{SiC}}{\sigma_{SiC}} + \frac{2}{100}\frac{\Delta\sigma_{sub}}{\sigma_{sub}} \quad (5.3)$$

A reference resistor was also fabricated to investigate the relative conductance change of the substrate during bending experiment. At an applied strain of 800 ppm on the SiC resistor, the relative conductance change of the substrate was found to be below 0.1%. As a result, from Eqs. 5.1 and 5.3, the substrate only contributed less than 0.2% to the measured gauge factor, which is negligible. Thus, the gauge factor of p-type nc-SiC was found to be 14.5, which is much larger than that in most metals [15–19].

5.3 Discussion

The piezoresistive effect in p-type nc-SiC can be explained using the model of current transport in several polycrystalline and nanocrystalline materials [20–22], in which p-type nc-SiC contains several crystalline grains, connected to each other by grain boundaries, as shown in Fig. 5.4. For the sake of simplification, the structure of nc-SiC was considered in one dimension (1-D), as shown in Fig. 5.4. Based on this model, the total resistance of nc-SiC is the sum of grains' resistance (R_c) and resistance of boundary (R_b):

$$R_{nc-SiC} = R_c + R_b \quad (5.4)$$

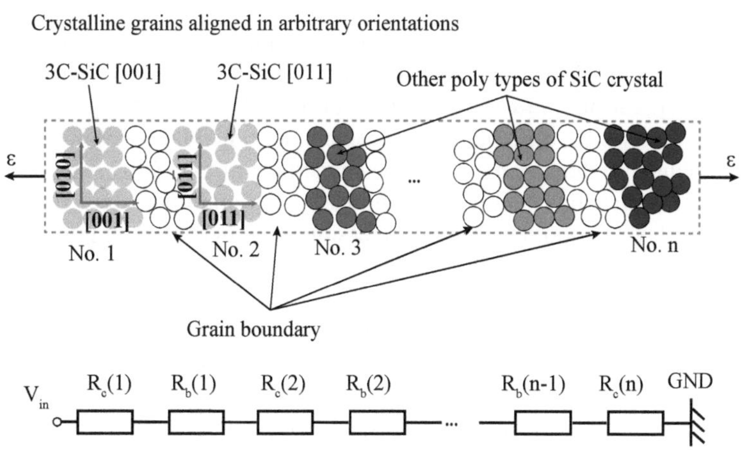

Fig. 5.4 A one dimensional (1-D) model of the crystalline structure of nc-SiC which contains a large number of crystalline SiC grains and grain-boundary. Reprint with permission from [6]

When a strain is applied, both resistance of grains and boundaries change, causing the change of the total resistance. Firstly, the piezoresistive effect in crystalline SiC grains was considered. Based on the deformation potential model, the piezoresistive effect in p-type semiconductors is due to the splitting of heavy holes and light holes in the valence band under applied strains [15, 23]. This band splitting leads to the redistribution of heavy holes and light holes in these two bands, which follows the rule that holes will fill lower energy levels. The re-population of holes results in the change of electrical resistivity of the crystalline SiC.

Secondly, the effect of strain on the resistance of grain boundaries was taken into account. The resistance of boundaries can be explained based on the tunneling current between neighbor grains. At grain boundaries, atoms have twisted/strained and/or dangling bonds. These defects result in the formation of trapping states which immobilize charge carriers, creating a potential barrier (V_b) [20, 21]. Consequentially, the carriers can move through these potential barriers via the tunneling current which is described in the following equation [24, 25]:

$$J \propto p_{hh} J_{hh} + p_{lh} J_{lh} \tag{5.5}$$

where p_{hh} and p_{lh} are the concentration of heavy hole and light hole; and J_{hh} and J_{lh} are tunneling current components of heavy hole and light hole which depend on the potential barrier ($\propto e^{qV_b/kT}$, where q is electron charge; V_b is the potential barrier; k is the Boltzmann constant; and T is the absolute temperature). Under strain, heavy holes and light holes concentrations change. As a consequence, the tunneling current varies with applied strain. Thus, the piezoresistive effect of nc-SiC grain boundaries in fact is established from the piezoresistive effect of each nc-SiC single crystal.

It is well-known that the piezoresistive effect in single crystalline SiC depends on crystallography orientations. As such, according to Sect. 4.2.2 in Chap. 4, regarding the orientation dependence of piezoresistive effect in p-type 3C-SiC, a large longitudinal gauge factor of 30.3 in [110] orientation, and a smaller longitudinal gauge factor of 5.0 in [100] orientation of single crystalline 3C-SiC, were reported [9]. Therefore, if the applied strain does not align with the orientation corresponding to the most significant piezoresistive effect of a single crystal, the gauge factor is smaller. In other words, the random arrangement of crystalline grains in nc-SiC may have contributed to the diminution of the gauge factor of nc-SiC in comparison to that of single crystalline SiC [21]. Additionally, the grain boundary scattering is expected to play an important role in decreasing the gauge factor of polycrystalline and nanocrystalline materials [26].

In conclusion, in this chapter, the piezoresistive effect of p-type nanocrystalline SiC grown by the LPCVD process was characterized. At a carrier concentration of $2 \times 10^{18}\,cm^{-3}$, the gauge factor of p-type nc-SiC was found to be 14.5, which is larger than metals, but smaller than single crystalline SiC. This result was explained due to the change of the resistivity of crystal grains and boundaries as well as the randomly aligned orientations of SiC crystals. Due to the capability of growing on various substrate (e.g. Si, SiO_2) at a lower temperature than single crystalline SiC, the significant piezoresistive effect in nc-SiC makes it a good candidate for MEMS mechanical sensors used in harsh environments and bio applications. Additionally, the insight of the piezoresistive effect in nc-SiC is also valuable for understanding the impact of mechanical strain on nc-SiC based devices such as solar cells which are affected by various types of stress during the fabrication process and in applications.

References

1. Y. Komura, A. Tabata, T. Narita, M. Kanaya, A. Kondo, T. Mizutani, Film properties of nanocrystalline 3C-SiC thin films deposited on glass substrates by hot-wire chemical vapor deposition using CH_4 as a carbon source. Jpn. J. Appl. Phys. **46**(1), 45–50 (2007)
2. B. Somogyi, V. Zolyomi, A. Gali, Near-infrared luminescent cubic silicon carbide nanocrystals for in vivo biomarker applications: an ab initio study. Nanoscale **4**, 7720–7726 (2012)
3. M. Eickhoff, M. Moller, G. Kroetz, M. Stutzmann, Piezoresistive properties of single crystalline, polycrystalline, and nanocrystalline n-type 3C-SiC. J. Appl. Phys. **96**, 2872–2879 (2004)
4. N. Naderi, M.R. Hashim, Nanocrystalline SiC sputtered on porous silicon substrate after annealing. Mater. Lett. **97**, 90–92 (2013)
5. L. Wang, S. Dimitrijev, P. Tanner, J. Zou, Aluminum induced in situ crystallization of amorphous SiC. Appl. Phys. Lett. **94**, 181909 (2009)
6. H.-P. Phan, D.V. Dao, L. Wang, T. Dinh, N.-T. Nguyen, A. Qamar, P. Tanner, S. Dimitrijev, Y. Zhu, The effect of strain on the electrical conductance of p-type nanocrystalline silicon carbide thin films. J. Mater. Chem. C **3**, 1172–1176 (2015)
7. R. Verucchi, L. Aversa, M.V. Nardi, S. Taioli, S. Beccara, D. Alfe, L. Nasi, F. Rossi, G. Salviati, S. Iannotta, Epitaxy of nanocrystalline silicon carbide on Si (111) at room temperature. J. Am. Chem. Soc. **2012**(134), 17400–17403 (2012)
8. H. Tateyama, H. Noma, Y. Adachi, M. Komatsu, Prediction of stacking faults in β-silicon carbide: X-ray and NMR studies. Chem. Mater. **9**(3), 766–772 (1997)
9. H.P. Phan, D.V. Dao, P. Tanner, N.T. Nguyen, L. Wang, Y. Zhu, S. Dimitrijev, Fundamental piezoresistive coefficients of p-type single crystalline 3C-SiC. Appl. Phys. Lett. **104**, 111905 (2014)
10. H.P. Phan, D.V. Dao, P. Tanner, N.T. Nguyen, J.S. Han, S. Dimitrijev, G. Walker, L. Wang, Y. Zhu, Thickness dependence of the piezoresistive effect in p-type single crystalline 3C-SiC nanothin films. J. Matter. Chem. C **2**, 7176–7179 (2014)
11. H.P. Phan, P. Tanner, D.V. Dao, N.T. Nguyen, L. Wang, Y. Zhu, S. Dimitrijev, Piezoresistive effect of p-type single crystalline 3C-SiC thin film. IEEE Electron Device Lett. **35**(3), 399–401 (2014)
12. A. Qamar, P. Tanner, D.V. Dao, H.P. Phan, T. Dinh, Electrical properties of p-type 3C-SiC/Si heterojunction diode under mechanical stress. IEEE Electron Device Lett. **35**(12), 1293–1295 (2014)
13. M.-D. Nguyen, H.-P. Phan, K. Matsumoto, I. Shimoyama, A sensitive liquid-cantilever diaphragm for pressure sensor, in *Proceedings of 26th IEEE International Conference on Micro Electro Mechanical Systems (MEMS 2013)* (2013)
14. D.V. Dao, T. Toriyama, J. Wells, S. Sugiyama, Six-degree of freedom micro force-moment sensor for application in geophysics, in *Proceedings of 15th IEEE International Conference on Micro Electro Mechanical Systems (MEMS 2002)*, Las Vegas, USA (2002), pp. 20–24
15. A.A. Barlian, W.T. Park, J.R. Mallon Jr., A.J. Rastegar, B.L. Pruitt, Review: semiconductor piezoresistance for microsystems. Proc. IEEE **97**(3), 513 (2009)
16. J. Engel, J. Chen, C. Liu, Development of polyimide flexible tactile sensor skin. J. Micromech. Microeng. **13**, 359–366 (2003)
17. W.K. Schomburg, Z. Rummler, P. Shao, K. Wulff, L. Xie, The design of metal strain gauges on diaphragms. J. Micromech. Microeng. **14**, 1101–1108 (2004)

18. I.H. Kazi, P.M. Wild, T.N. Moore, M. Sayer, Characterization of sputtered nichrome (Ni-Cr 80/20 wt.%) films for strain gauge applications. Thin Solid Films **515**(4), 2602–2606 (2006)
19. D. Macnamara, D. Thiel, D. James, P. Lisner, in *Proceedings of SPIE 6035, Microelectronics: Design, Technology, and Packaging II*, ed. by A.J. Hariz (SPIE, 2006), pp. 603524
20. V. Mosser, J. Suski, J. Goss, E. Obermeier, Piezoresistive pressure sensors based on polycrystalline silicon. Sens. Actuators A Phys. **28**(2), 113–132 (1991)
21. P. Alpuim, M. Andrade, V. Sencadas, M. Ribeiro, S.A. Filonovich, S. Lanceros-Mendez, Piezoresistive properties of nanocrystalline silicon thin films deposited on plastic substrates by hot-wire chemical vapor deposition. Thin Solid Films **515**, 7685–7661 (2007)
22. S.D. Janssens, S. Drijkoningen, K. Haenen, Large piezoresistive effect in surface conductive nanocrystalline diamond. Appl. Phys. Lett. **105**, 101601 (2014)
23. K. Nakamura, D.V. Dao, Y. Isono, T. Toriyama, S. Sugiyama, Nanowires, *Electronic States and Piezoresistivity in Silicon Nanowires* (In-Tech, Vienna, 2010)
24. X. Liu, C. Shi, R. Chuai, Polycrystalline silicon piezoresistive nano thin film technology, *Solid State Circuits Technologies* (2010). ISBN: 978-953-307-045-2
25. C. Chuai, B. Liu, X. Liu, X. Sun, X. Shi, L. Wang, Design, fabrication and characterization of a high-sensitivity pressure sensor based on nano-polysilicon thin film transistors. J. Semicond. **31**(3), 032002 (2010)
26. L. Fang, W.L. Wang, P.D. Ding, K.J. Liao, J. Wang, Study on the piezoresistive effect of crystalline and polycrystalline diamond under uniaxial strains. J. Appl. Phys. **86**(9), 5185–5193 (1999)

Chapter 6
The Piezoresistive Effect of Top Down p-Type 3C-SiC Nanowires

The piezoresistance of SiC nanowires is of interest as a means to scale down devices size as well as to enhance the sensitivity of sensors [1–3]. However, as presented in Chap. 1, there have been a limited number of experimental work on the piezoresistance of SiC, which were fabricated using bottom up process [2–4]. This chapter presents the fabrication of SiC nanowires using a top down process. Two configurations which are non-released and suspended SiC nanowires were formed, and a comparison between the gauge factors of these nanowires with their bulk SiC counterparts were also discussed.

6.1 Piezoresistance of Non-released SiC Nanowires

6.1.1 Fabrication of Top Down SiC Nanowires

Figure 6.1a present the top down fabrication process of nanowires. Initially p-type I-shaped SiC resistors was formed from a single crystal 3C-SiC on Si wafer using the micro lithography described in Chap. 3. The carrier concentration of the SiC resistors were similar to the samples reported in the previous chapters, which was approximately $5 \times 10^{18}\,\text{cm}^{-}3$. Next, SiC on Si beams with dimension of 60 mm in length, 8 mm in width, and 625 µm in thickness were cut from the patterned wafer. Silicon carbide nanowires were then fabricated employing gallium ion (Ga^{+}) FIB (™HITACHI FB 2200) [5] at a applied voltage of 40 kV and an DC current of 0.07 nA. Figure 6.1b shows the colorized SEM image of a 3C-SiC nanowires array. Three nanowires was formed on each array since it enhance the conductance of SiC nanowire resistors, thus ease the measurement. The dimensions of each SiC nanowires were 5 µm in length, 300 nm in width and 300 nm in thickness. In addition,

H.-P. Phan, *Piezoresistive Effect of p-Type Single Crystalline 3C-SiC*,
Springer Theses, DOI 10.1007/978-3-319-55544-7_6

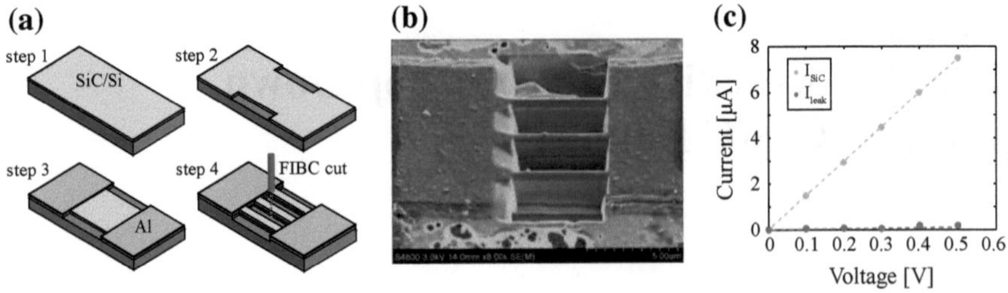

Fig. 6.1 **a** Top down fabrication of non released SiC nanowires; **b** Colorized SEM image of SiC NWs. **c** The I-V curve of a SiC nanowire array in comparison to the current leakage from SiC to the Si substrate

the orientation of SiC nanowires were [110], as this direction offers a higher gauge factors compared to other direction in (100) plane.

6.1.2 The Gauge Factor of the Non-released SiC Nanowires

The I-V of the SiC nanowires was obtained using an Agilent 2722A™ power source, showing a linear relationship resulted from the good Ohmic contact formed between Al and Si, as shown in Fig. 6.1c. Furthermore, the leakage current to the substrate was explored using the same technique described in Chap. 4. Accordingly, a small leakage of 20 nA was observed at an applied voltage of 0.5 V, which can be neglected in comparison to that supplied to the SiC nanowires.

The strains induced into the SiC nanowires by bending the SiC/Si beam were estimated using finite element analysis. The simulation data obtained using COMSOL Multiphysics™ showed that more than 98% of the strain applied to the Si substrate was induced into the SiC nanowires, Fig. 6.2b. The simulation also showed a slightly higher strain at the area connecting the SiC nanowires and electrode pads. Nevertheless, this strain did not considerably modify the average strain applied to the SiC nanowires, since it only located at a relatively small area. Therefore, when deflecting the free end of the SiC/Si beam with a force ranging from 0 to 0.4 N, the strain applied to the SiC nanowires varied from 0 to 280 ppm.

The resistance change of the SiC nanowire under strain was converted into an output voltage and then amplified using the circuit illustrated in Fig. 6.3a. The resistance change ($\Delta R/R$) can be obtained from the measured output (V_o) using the following equation:

$$V_o = \frac{V_i}{4}\frac{\Delta R}{R}G \tag{6.1}$$

where $V_i = 1V$ was input voltage of the Wheatstone bridge, and $G = 400$ was the voltage amplification gain.

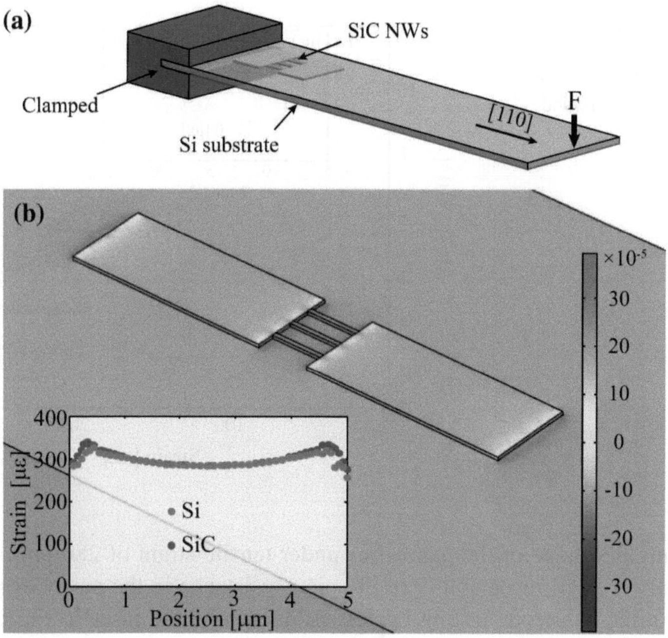

Fig. 6.2 Simulations the strain induced into SiC nanowires using finite element analysis. Inset: A comparison between the strain applied to the SiC nanowires and that of the top surface of the Si substrate

Fig. 6.3 **a** The readout circuit; **b** The output voltage of a SiC nanowires when applying a tensile strain of 280 ppm for three cycles

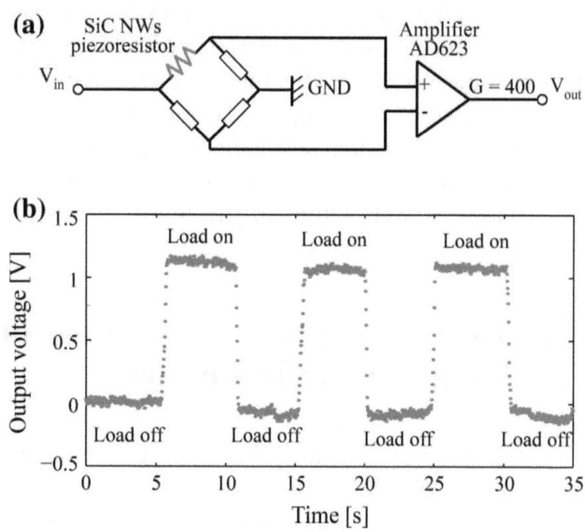

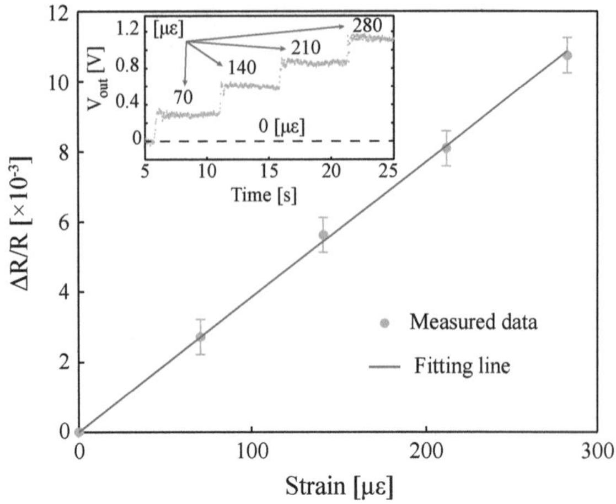

Fig. 6.4 The relationship between the resistance change ($\Delta R/R$) of the SiC nanowire and the applied strain

The output voltage of the nanowires under tensile strain of 280 ppm is plotted in Fig. 6.3b. A good repeatability of the piezoresistance in the nanowires was also observed after numerous testing cycles. Additionally, the inset in Fig. 6.4 shows that increasing tensile strains leaded to an increase in the output voltage of the SiC nanowires. Based on this data, $\Delta R/R$ was obtained and plotted against the tensile strains, as shown in Fig. 6.4. A clear linear behavior of the resistance change in respect of the applied tensile strain was also confirmed. The gauge factor the top down nanowires was calculated at 35 [6].

Compared to the gauge factor of bulk SiC presented in the previous chapter, the top down non-released nanowires show a comparable piezoresistance. Even though the giant piezoresistive effect was not observed in the as-fabricated nanowires, the proposed design can be employed to miniaturize SiC mechanical sensors. In addition, since the SiC nanowires was fabricated from a large scale SiC/Si wafer, it can also be integrated with other SiC electrónic devices. This is a clear advantage of top down SiC nanowires over bottom up growth SiC nanowires.

6.2 The Piezoresistance of Suspended SiC Nanowire Using Nano-strain Amplifier

As presented in the section above, the gauge factor of the piezoresistance in non released top down SiC nanowires was almost similar to that of the bulk SiC, and non giant piezoresistance was observed. This is considerable since the smallest dimension of the nanowires was relative large where the quantum confinement is in effect. However, scaling down the nanowire using the top down technique is a challenging task as the FIB could induced significant damage layer in extremely thin or narrow nanowires. This section presents an innovative mechanical approach where top down suspended nanowires is utilized to significantly improve the sensitivity of SiC piezoresistive strain sensors. The developed structure (namely nano-strain amplifier) can also be applied in other materials [7].

6.2.1 The Principle of the Nano-strain Amplifier

The sensitivity (S) of a strain sensors is defined based on the ratio of relative resistance change ($\Delta R/R$) to the strain applied to the substrate (ε_{sub}): $S = (\Delta R/R)/\varepsilon_{sub}$. Therefore, for a given applied strain, to obtain a high sensitivity, a structure or a material which offers high resistance change ($\Delta R/R$) is desired. In addition, $\Delta R/R$ is expressed through the gauge factor (GF) and the strain induced into the nanowires (ε_{ind}): $\Delta R/R = GF\varepsilon_{ind}$. In fact, the piezoresistance of SiC nanowires on Si with the smallest feature above 300 nm was characterized in the previous section, showing a comparable gauge factor with bulk Si. Consequently, to improve the sensitivity of SiC nanowire strain sensors ($\Delta R/R$), amplifying the strain of SiC nanowire ε_{ind} is a more feasible approach.

The concept of the conventional nanowire piezoresistors are illustrated in Fig. 6.5a, b, where the nanowires can be kept on the substrate or suspended. From the simulation data presented in the previous sections, the strain induced into the non-released SiC nanowire or suspended nanowire almost equal the strain applied to the substrate. Therefore, the use of conventional structure is not suitable to induce a large strain into SiC nanowires. A schematic sketch of the proposed structure to amplify the strain of SiC nanowires is presented in Fig. 6.5c. In this structure, the nanowires are locally fabricated at the centre of a suspended micro frame which was released from the Si substrate. The key principle of this nano strain-amplifier is similar to the well-known dogbone structure where strains or stresses can be concentrated at the narrow area [8, 9].

The model of the nano strain-amplifier is simplifier as illustrated in Fig. 6.5d, where the nanowires and micro frames are considered as serially connected springs. In addition, since the stiffness of a spring is proportional to its width (w), and inversely proportional to its length (l), the spring corresponding to the SiC nanowires is softer than that of the micro frames. Therefore, when a tensile strain is applied to this serial

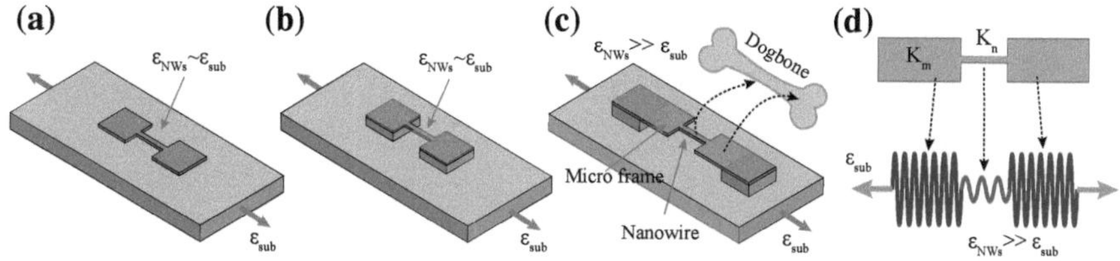

Fig. 6.5 Schematic sketches of nanowire strain sensors. **a, b** Conventional non-released and released NW structure; **c, d** The nano strain-amplifier and its equivalent spring model

spring system, the softer spring will be significantly elongated in comparison to the stiffer one [10].

The strain-amplification gain can be quantified as follows. Let L, L_m, L_n be the lengths of the suspended structure, the micro frames, and the nanowires, respectively. The total displacement of the suspended structure is given by:

$$\Delta L = \Delta L_m + \Delta L_n = L_m \varepsilon_m + L_n \varepsilon_n \tag{6.2}$$

where ε_m, ε_n are the strains induced into the micro spring and nano spring, respectively. In addition, the equilibrium of the stressing force (F) along the series of springs results in the following relationship:

$$F = K_m \Delta L_m = K_n \Delta L_n \tag{6.3}$$

where K_m and K_n are the stiffness of the suspended micro frames and nanowires, respectively. Equation 6.3 can be re-written as:

$$\frac{\Delta L_m}{\Delta L_n} = \frac{K_n}{K_m} = \frac{L_m w_n}{L_n w_m} \tag{6.4}$$

Consequently, the displacement of the SiC nanowires is:

$$\Delta L_n = \Delta L \frac{\omega_m L_n}{\omega_m L_n + \omega_n L_m} = \Delta L \frac{L_n}{L_n + \frac{\omega_n}{\omega_m} L_m} \tag{6.5}$$

From Eqs. 6.4 and 6.2, the strain applied to the nanowires located at the centre of the suspended frame is:

$$\begin{aligned} \varepsilon_n = \frac{\Delta L_n}{L_n} &= \frac{1}{L_n + \frac{\omega_n}{\omega_m} L_m} \Delta L \\ &= \frac{1}{\frac{L_n}{L} + \frac{\omega_n}{\omega_m} \frac{L_m}{L}} \frac{\Delta L}{L} \\ &= \frac{1}{1 - \frac{L_m}{L} + \frac{\omega_n}{\omega_m} \frac{L_m}{L}} \varepsilon_{sub} \\ &= \frac{1}{1 - \frac{\omega_m - \omega_n}{\omega_m} \frac{L_m}{L}} \varepsilon_{sub} \end{aligned} \tag{6.6}$$

Evidently, from Eq. 6.6, increasing the ratio of w_m/w_n and L_m/L_n can significantly enlarge the strain (ε_n) of the SiC nanowire.

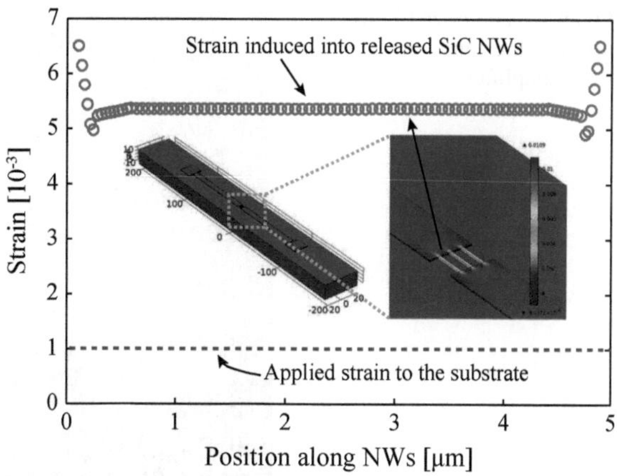

Fig. 6.6 Simulation results of the strain induced in to a SiC nanowire array with the width of each wire being 380 nm

A simulation using ™COMSOL Multiphysics was also carried out to verify using the numerical analysis. In the finite element analysis, the width of the suspended frame was set to be 8 μm, whereas that of each nanowire was 370 nm. Additionally, the ratio of the length of nanowires and micro bridge was set to be 1: 20. Furthermore, three nanowires were formed on each array, since they offer a better electrical conductance than a single wire, making the subsequent experimental demonstration easier. The simulation data showed that the strain applied to the SiC nanowires in the strain-amplifier was approximately 5.9 times larger than that of the substrate, Fig. 6.6. This result was consistent with that estimated by using Eq. 6.6. A simulation was also performed on a nanowire array with an average width of 470 nm, showing an strain-amplification gain of 4.5.

6.2.2 Demonstration of Nano Strain-Amplifier in p-Type 3C-SiC Nanowires

The nano strain-amplifier used to enhance the sensitivity of SiC strain sensors was demonstrated in the following experiments. The fabrication previous was similar to that reported in the previous sections, where SiC nanowires were formed on a suspended p-type SiC bridge using FIB. The longitudinal orientation of the nanowires was [110], and its carrier concentration was 5×10^{18} cm^{-3}. Two types of nanowire array were fabricated in which the width of each nanowire in each type was 380 and 470 nm, respectively. Figure 6.7 shows the colorized SEM images of SiC nanowires fabricated at the centre of a SiC micro bridge.

The bending experiment was then employed where the strain induced into the Si substrate was estimated utilizing the beam model presented in Chap. 4. Accordingly, the strain applied to the Si substrate ranged from 0 to 2×10^{-4} ppm. Figure 6.8 shows a linear relationship between the resistance change of the SiC nanowires ($\Delta R/R$) and

Fig. 6.7 A colorized SEM image of the nano strain-amplifier

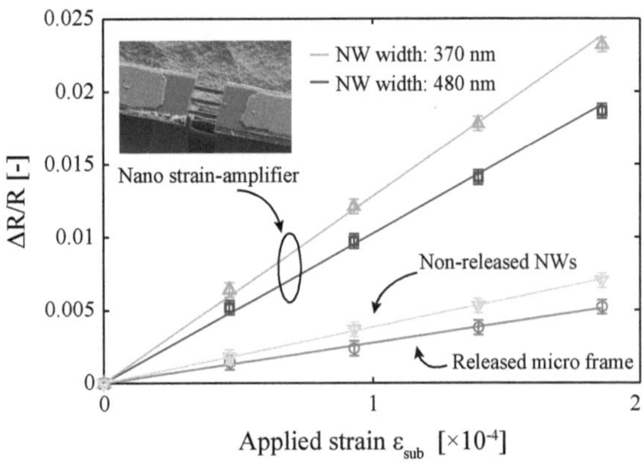

Fig. 6.8 A comparison between the resistance change of the nano strain-amplifier and the conventional structure under the same strain applied to the Si substrate

the strain of the Si substrate. To make a comparison with the conventional structures, the resistance change of non-released SiC nanowires and released micro frames were also plotted on the same graph. From the resistance change of SiC in accordance to the strain applied to the substrate, the effective gauge factor can be estimated as: $GF_{eff} = (\Delta R/R)/\varepsilon_{sub}$. As reported in Chap. 4, the effective gauge factor of the suspended micro SiC frame was 28, while that of the non-released SiC nanowires was 35. The effective gauge factor of the 380 and 470 nm SiC nanowires in the nano strain-amplifier were found to be 150 and 124, respectively, which are approximately 5.4 times and 4.6 times larger than the conventional structures. Therefore, the proposed strain-amplification structures is an excellent platform for the development of ultra sensitive strain sensors.

References

1. H.-P. Phan, D.V. Dao, K. Nakamura, S. Dimitrijev, N.-T. Nguyen, The piezoresistive effect of SiC for MEMS sensors at high temperatures: a review. J. Microelectromech. Syst. **24**(6), 1663–1677 (2015)
2. R. Shao, K. Zheng, Y. Zhang, Y. Li, Z. Zhang, X. Han, Piezoresistance behaviors of ultra-strained SiC nanowires. Appl. Phys. Lett. **101**(23), 233109 (2012)
3. F. Gao, J. Zheng, M. Wang, G. Wei, W. Yang, Piezoresistance behaviors of p-type 6 H-SiC nanowires. Chem. Comm. **47**(43), 11993–11995 (2011)
4. J. Bi, G. Wei, L. Wang, F. Gao, J. Zheng, B. Tang, W. Yang, Highly sensitive piezoresistance behaviors of n-type 3C-SiC nanowires. J. Mater. Chem. C **1**(30), 4514–4517 (2013)
5. T. Dinh, H.-P. Phan, T. Kozeki, A. Qamar, T. Namazu, N.-T. Nguyen, D.V. Dao, Thermoresistive properties of p-type 3C-SiC nanoscale thin films for high-temperature MEMS thermal-based sensors. RSC Adv. **5**, 106083–106086 (2015)
6. H.-P. Phan, T. Dinh, T. Kozeki, T.-K. Nguyen, A. Qamar, T. Namazu, N.-T. Nguyen, D.V. Dao, The piezoresistive effect in top-down fabricated p-type 3C-SiC nanowires. IEEE Electron Device Lett. **37**(8), 1029–1032 (2016)
7. H.-P. Phan, T. Dinh, T. Kozeki, T.-K. Nguyen, A. Qamar, T. Namazu, N.-T. Nguyen, D.V. Dao, Nano strain-amplifier: Making ultra-sensitive piezoresistance in nanowires possible without the need of quantum and surface charge effects. Appl. Phys. Lett. **109**(12), 123502 (2016)
8. H. Huang, F. Spaepen, Tensile testing of free-standing Cu, Ag and Al thin films and Ag/Cu multilayers. Acta Materialia **48**(12), 3261–3269 (2000)
9. J.M. Wernik, S.A. Meguid, On the mechanical characterization of carbon nanotube reinforced epoxy adhesives. Mater. Des. **59**, 19–32 (2014)
10. R.C. Hibbeler, *Engineering Mechanics: Statics*, 14th edn. (Pearson, 2015)

Chapter 7
Conclusion and Perspectives

7.1 Conclusion

This dissertation presented the piezoresistive effect in p-type single crystalline 3C-SiC, including the gauge factor, the piezoresistive coefficients, and its thickness–, orientation–, and temperature– dependence. The piezoresistive effect in p-type 3C-SiC four-terminal resistors was also presented. Furthermore, a comparison between the piezoresistive effect of p-type single crystalline and p-type nanocrystalline SiC was also reported.

In summary, through this search, the following objectives have been successfully addressed:

(I) The gauge factor in [110] orientation of the p-type single crystalline 3C-SiC films was measured to be approximately 30. The gauge factor found in p-type 3C-SiC was larger than that of n-type 3C-SiC reported by Shor et al., with the same range of carrier concentration ($10^{18} \sim 10^{19}$ cm^{-3}). This result indicates the advantage of p-type 3C-SiC over n-type 3C-SiC. Additionally, with the advantages of 3C-SiC such as wide band gap, high melting point, high Q factor, excellent chemical inertness, as well as the ability of epitaxial growth on a large scale Si substrate, the piezoresistive effect in p-type single crystalline 3C-SiC is a potential candidate for high temperature, high frequency, and bio compatible applications.

(II) The fundamental piezoresistive coefficients of p-type single crystalline 3C-SiC in the principal coordinate π_{11}, π_{12} and π_{44} were found to be 1.5×10^{-11}, -1.4×10^{-11} and 18.1×10^{-11} Pa^{-1}, respectively, indicating that the gauge factor of p-type 3C-SiC resistor is significantly dominated by the shear piezoresistive coefficient π_{44}. Additionally, from the graphical study on the dependence of the piezoresistive coefficients on orientation, it is suggested that, when designing p-type 3C-SiC based mechanical sensors in the (100) plane, the [110] orientation should be selected to achieve maximum sensitivity.

(III) The thickness dependence of the piezoresistive effect in p-type 3C-SiC were characterized by investigating the gauge factor in 3C-SiC films with thicknesses

H.-P. Phan, *Piezoresistive Effect of p-Type Single Crystalline 3C-SiC*,
Springer Theses, DOI 10.1007/978-3-319-55544-7_7

varying from 80 nm to 1 μm. The consistency of the gauge factor of the SiC layer with a thickness above 280 nm, and the large drop of the gauge factor in the 80 and 130 nm films imply that the crystal defect has a significant influence on the piezoresistive effect of the p-type single crystalline 3C-SiC nano thin film with thickness below 150 nm, while this influence is negligible in sufficiently thick film. These results suggest that, when designing MEMS mechanical sensors utilizing the piezoresistive effect of ultra thin 3C-SiC films, the effect of the crystal defect distributed at the SiC/Si interface should be taken into account.

(IV) The piezoresistive effect of p-type 3C-SiC at high temperature was characterized using an in situ experimental method. The stability of the gauge factor of p-type 3C-SiC with a large temperature range from 300 to 600 K, demonstrated the feasibility of this poly type for mechanical sensing at high temperatures. In addition, a coupling phenomenon of the piezoresistive effect and thermoresistive effect on locally heated SiC resistors under strain was also reported. The obtained results indicated that by utilizing this coupling effect, it is possible to enhance the gauge factor of 3C-SiC by 20% at approximately 600 K. Furthermore, the in situ method proposed in this work is also applicable for characterizing the piezoresistive effect of other semiconductor materials grown on an electrical conductive material.

(V) Experimental results obtained in the piezoresistive effect in 3C-SiC four-terminal devices showed a large voltage at a certain applied current and stress. This large generated voltage along with a direct/simple signal-readout method indicate that 3C-SiC four-terminal resistors are applicable for MEMS mechanical sensors. Additionally, based on the shear piezoresistive coefficient π_{61}, π_{62}, and π_{66} found in this work, its is possible to predict the piezoresistive effect in four-terminal SiC resistors fabricated on (100) plane.

(VI) The piezoresistive effect in p-type nanocrystalline SiC was also investigated, showing a gauge factor of approximately 14, which is about 5 times larger than that of metals, but 2 times smaller than single crystalline p-type 3C-SiC. The diminution of the piezoresistive effect in nc-SiC compared to crystalline SiC was considered due to the randomly oriented crystalline grains within nc-SiC films, and the influence of the grain boundaries. However, as nc-SiC can be directly grown and on a insulation substrate (e.g. glass), the piezoresistive effect in nc-SiC can be used for sensors operating at high temperatures.

VII P-type SiC nanowires were patterned using a top down process, where FIB was utilized to form nano scale structures. The gauge factor of non-released 3C-SiC nanowires was found to be 35, which was almost in the same range as that of bulk SiC. To improve the sensitivity of SiC strain sensors, a nano strain-amplifier was developed where SiC nanowires were locally formed on a suspended SiC micro frame. The effective gauge factor of the SiC nanowires with a width of 380 nm were obtained to be approximately 150. This demonstrated the potential of using nano strain-amplifier to develop ultra sensitive NEMS mechanical sensors.

7.2 Research Perspectives

Utilizing the piezoresistive effect of the p-type 3C-SiC, it is possible to develop sensors used for high temperature applications such as strain sensors and pressure sensors. To make these sensors, an electrical insulation is required to prevent the leakage current to the Si substrate at elevated temperatures. Consequently, the author suggest wafer bonding techniques which have been developed by numerous research groups to produce p-type 3C-SiC on insulator platforms. In addition, owing to the excellent mechanical properties and chemical inertness of SiC, the use of the piezoresistive effect in p-type 3C-SiC could extend beyond sensors operating at high temperatures to cover high-frequency and bio applications. For instance, utilizing the large gauge factor of p-type SiC at room temperature could make the development of SiC based self-sensing high-Q-factor/high-frequency resonators possible, and thereby may eliminate the need for optical measurements. On the other hand, the piezoresistive effect of SiC can also be employed to characterize the mechanical properties of cells.

Further investigations on the piezoresistive effect in p-type 3C-SiC can be carried out towards the miniaturization of SiC nanostructures such as nanowires or nano thin film Field Effect Transistor (FET). When the dimensions of SiC nano structures are shrank down (e.g. with the smallest dimension below 10 nm), where the quantum confinement becomes effective, the piezoresistive effect in 3C-SiC is expected to be significantly changed.

Appendix A
Process Flow of the SiC/Si Beam

See Tables A.1 and A.2.

Figures A.1, A.2, and A.3 show the design of the Cr masks used to fabricate the U-shape SiC resistors, released resistors, and four-terminal devices, respectively.

Table A.1 Fabrication process

Step	Task	Process
1	Wafer preparation	Growth SiC on Si
2	Fabricate SiC resistor	Spincoat photoresist (AZ6612)
		Bake photoresist at 110 °C for 70 s
		Align the first mask
		Post bake the photoresist at 126 °C for 70 s
		Photoresist development
		Plasma etching SiC by SF_6 gas
		Remove the photoresist by acetone and IPA
3	Patterning electrode	Deposit aluminum by sputtering (∼100 nm)
		Spincoat photoresist (AZ6612)
		Bake photoresist at 110 °C for 70 s
		Align the second mask
		Post bake the photoresist at 126 °C for 70 s
		Photoresist development
		Etching aluminum by Al etchant
		Remove the photoresist by acetone and IPA
4	Dicing SiC/Si strips	Spincoat photoresist for protection
		Dice SiC/Si wafer into smaller strips
		Remove the photoresist by acetone and IPA

H.-P. Phan, *Piezoresistive Effect of p-Type Single Crystalline 3C-SiC*,
Springer Theses, DOI 10.1007/978-3-319-55544-7

Table A.2 List of equipment used in this work

No.	Equipment	Make and model	Purpose
1	HMDS oven	YES 3/10	Provide HMDS treatment to enhance the adhesion between photoresist and films
2	Spin coater	SSE OPTIcoat ST22+	Precisely coat photoresist
3	Softbake	–	For pre-bake and post-bake photoresist
4	Mask aligner	Quintel Ultra μ-line7000	Align patterns and exposure photoresist
5	Resist development	SSE OPTIcoat ST22	Develop photoresist
6	Epitaxial growth	MkII SPTS-Epiflx R+D	Grow 3C-SiC films on Si wafers
7	Plasma etch	STS LPX ICP SR	Etch SiC
8	Resist strip	Tegal 915	Remove photoresist
9	Metal sputtering	Surrey Nano Systems-Gamma	Deposit Al
10	FIB	–	Transfer SiC resistor
10	Thin film measurement	Nanospec AFT 210	SiC thickness measurement
11	Atomic Force Microscope	Park AFM NX20	Characterization of SiC surface roughness
12	SEM	Jeol JSM 6510LV - 2009	Image of SiC films and devices
13	Optical microscope	Olympus MX50AF -IC inspection microscope	photographs of devices
14	Semiconductor parametric analyser	HP 4145B	IV measurement and current/voltage supply
15	Multimeter	Agilent 34410A Multimeter	Voltage and resistance measurement
16	Power source	Agilent 2722A	Current supply of four-terminal resistances
17	Oscilloscope	Agilent MSO-X 3104A	Record the output voltage of piezoresistors
18	High temperature oven	–	Characterization of thermoresistive effect in SiC
19	Thermal chuck	–	For characterization of current leak through the SiC/Si junction
20	Thermo couple	K-type	For monitoring the temperature inside oven
21	Probe station	–	For fundamental electrical characterization
22	Wire bonding machine	Westbond 747630 E	For electrical contact
23	Xenon Di-fluoride (XeF_2) dry etcher	–	Isotropic Si etching

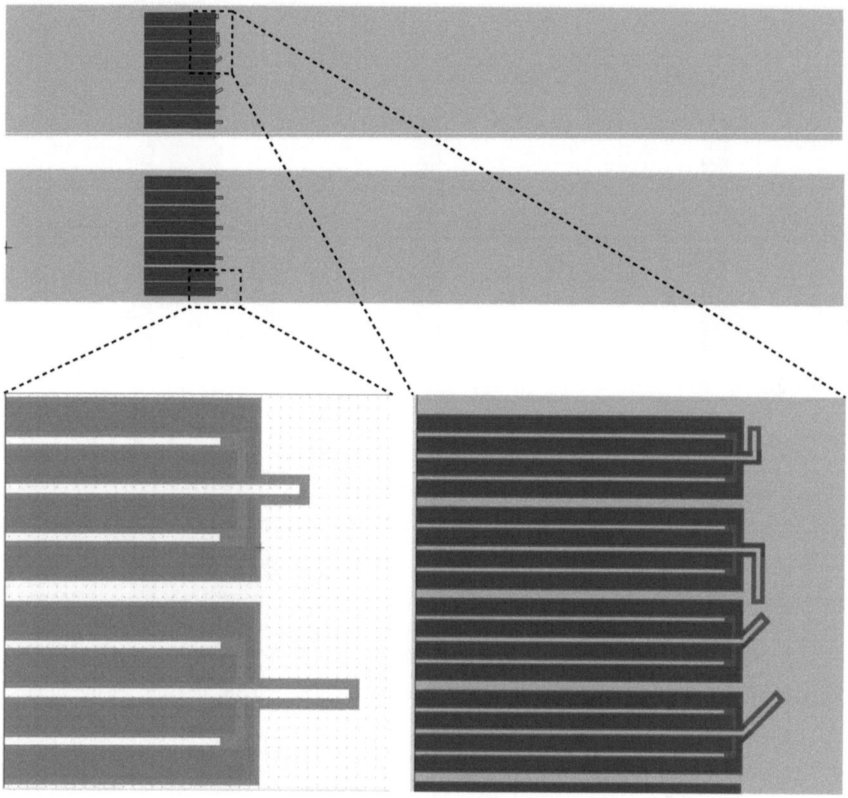

Fig. A.1 Mask design for the fabrication of SiC/Si cantilevers with U-shape SiC resistors on the surface

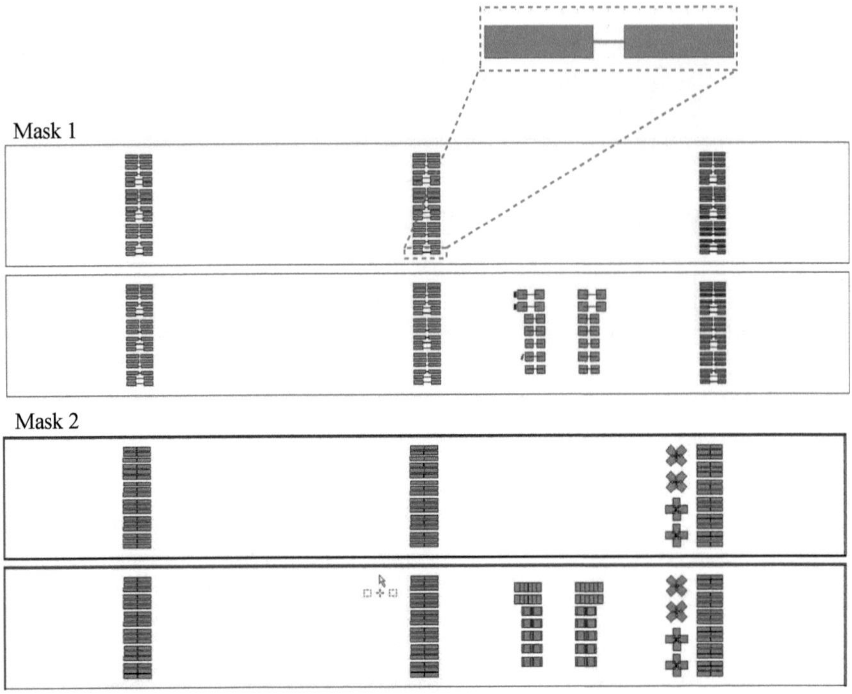

Fig. A.2 Mask design for the fabrication of released and transferred SiC resistors on the surface

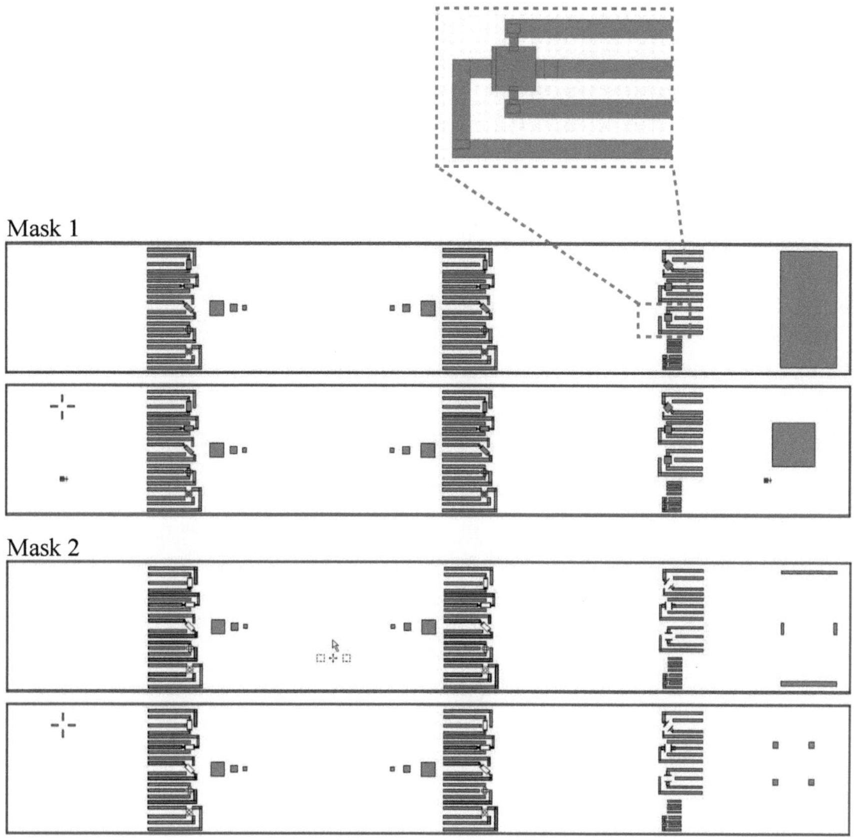

Fig. A.3 Mask design for the fabrication of SiC four-terminal resistors on the surface

Appendix B
Estimation of the Error Between the Simulation and the Actual Results of the Strain in 3C-SiC Resistor

To determine the strain in a bending beam problem, the common methods are using: (a) strain gauge embedded or bonded on the surface of the beam, (b) FEA (finite element analysis) simulation and (c) analytical calculation. While the strain measurement using strain gauge is convenient for large beam structures, the FEA (finite element analysis) simulation is suitable for estimating strain in small and complicated structures. The difference (error) between the simulation and the actual results, if happens, it should mainly come from the difference between the input parameters and the actual values. The input parameters include the dimensions of the Si beam, the misalignment of the SiC resistor on the Si beam, and the Youngs modulus.

When the thickness of SiC film is much smaller than that of the Si beam (the case of this work), the strain in the SiC resistor is approximately equal to the strain of Si at the SiC/Si interface and can be calculated as:

$$\varepsilon = \frac{6Fl}{Ewt^2} \tag{B.1}$$

where F is the applied force, and l, w, t are the length, width and thickness of the Si beam, respectively. The difference between the strain calculated from Eq. 4.2 (in Chap. 4) and Eq. B.1 written above is less than 0.3%. Assuming that the Young's modulus of Si substrate is known, the error from calculation is due to the error of the input parameters (F, l, w, and t) which can be estimated from:

$$\left|\frac{\Delta\varepsilon}{\varepsilon}\right| = \left|\frac{\Delta F}{F}\right| + \left|\frac{\Delta w}{w}\right| + \left|\frac{\Delta l}{l}\right| + 2\left|\frac{\Delta t}{t}\right| \tag{B.2}$$

The error of the applied force is caused by the error of the standard weight:

$$\left|\frac{\Delta F}{F}\right| = 2\% \tag{B.3}$$

Errors of width and length are caused by the dicing process (error = 10 μm):

H.-P. Phan, *Piezoresistive Effect of p-Type Single Crystalline 3C-SiC*, Springer Theses, DOI 10.1007/978-3-319-55544-7

$$|\frac{\Delta w}{w}| = \frac{10}{9000} \tag{B.4}$$

$$|\frac{\Delta l}{l}| = \frac{10}{60000} \tag{B.5}$$

Error of the thickness due to the reading error of micrometer (5 μm):

$$|\frac{\Delta t}{t}| = \frac{5}{625} \tag{B.6}$$

The angular error due to misalignment of SiC resistor on Si beam (about 2° according to the mask aligner specifications). Let θ be the angular error, the stress components of the beam will be [1]:

$$\begin{pmatrix} \sigma_{11} \\ \sigma_{22} \\ \sigma_{12} \end{pmatrix} = \begin{pmatrix} \cos^2\theta & \sin^2\theta & \sin 2\theta \\ \sin^2\theta & \cos^2\theta & -\sin 2\theta \\ -\frac{\sin 2\theta}{2} & \frac{\sin 2\theta}{2} & \cos 2\theta \end{pmatrix} \begin{pmatrix} \sigma \\ 0 \\ 0 \end{pmatrix} \tag{B.7}$$

Hence, the error of strain due to misalignment can be deduced as:

$$|\frac{\Delta \varepsilon}{\varepsilon}| = |\cos^2\theta - \gamma \sin^2\theta - 1| = (1+\gamma)\sin^2\theta \approx 0.002 \tag{B.8}$$

where $\gamma = 0.28$ is the Poisson's ratio of Si in [100] direction.

The error of Young's modulus caused by the dicing process. The dicing process may induce an angular error of about 2° in the orientation of Si beam which alters its Young's modulus. From the orientation dependence of Young's modulus of (100) Si reported in [2], we estimate the error Youngs modulus to be: $|\Delta E/E| \approx 1\%$.

Therefore, the total error of the estimated strain is about 5%.

References

1. C.-H. Cho, Alignment error analysis in resistor stress sensor on (001) silico. J. Korean Phys. Soc. **59**(3), 2209 (2011)
2. M.A. Hopcroft, W.D. Nix, T.W. Kenny, What is the Youngs modulus of Silicon. J. Microelectromechanical Syst. **19**(2), 229 (2010)

Appendix C
Estimation of the Piezoresistance in 3C-SiC Using the PZR Coefficients

The following experiment was carried out in order to compare the experimental results of the relative resistance change in p-type 3C-SiC resistors to the theoretical values estimated based on the three principle piezoresistive coefficients reported in Chap. 4. A general concept of a SiC resistor on (100) plane is described in Fig. C.1, in which the SiC resistor is aligned in an arbitrary orientation and the applied stress is also induced in a random direction. Let θ' be the angle between longitudinal axis of the SiC resistor and the direction of applied stress (Fig. C.2). Using Mohr's circle, the components of stress induced into the SiC resistors are:

$$\begin{cases} \sigma_l = \cos^2\theta' \cdot \sigma \\ \sigma_t = \sin^2\theta' \cdot \sigma \\ \sigma_s = -\frac{1}{2}\sin 2\theta' \cdot \sigma \end{cases} \tag{C.1}$$

Additionally, given that θ is the angle between longitudinal axis of the SiC resistor and [100] orientation in (100) plane, the piezoresistive coefficients π_l, π_t, π_s are deduced from the fundamental piezoresistive coefficients as:

$$\begin{cases} \pi_l = \pi_{11} - \frac{1}{2}(\pi_{11} - \pi_{12} - \pi_{44})\sin^2 2\theta \\ \pi_t = \pi_{12} + \frac{1}{2}(\pi_{11} - \pi_{12} - \pi_{44})\sin^2 2\theta \\ \pi_s = -\frac{1}{2}(\pi_{11} - \pi_{12} - \pi_{44})\sin 4\theta \end{cases} \tag{C.2}$$

Therefore, the relative resistivity change of a SiC resistor aligned in an arbitrary orientation in (100) Si beam is:

$$\Delta\rho/\rho = \left[\pi_{11}\cos^2\theta' + \pi_{12}\sin^2\theta' - \frac{1}{2}(\pi_{11} - \pi_{12} - \pi_{44})\sin 2(\theta - \theta')\right]\sigma \tag{C.3}$$

In the experiment, four SiC resistors were fabricated on a Si cantilever, in which the longitudinal direction of the Si beam was [110], which is the same as the direction of applied stress. Additionally, the angle between the SiC resistors and the longitudinal axis of the Si beam θ' were designed to be 0, $\pi/6$, $\pi/4$, and $\pi/2$.

H.-P. Phan, *Piezoresistive Effect of p-Type Single Crystalline 3C-SiC*, Springer Theses, DOI 10.1007/978-3-319-55544-7

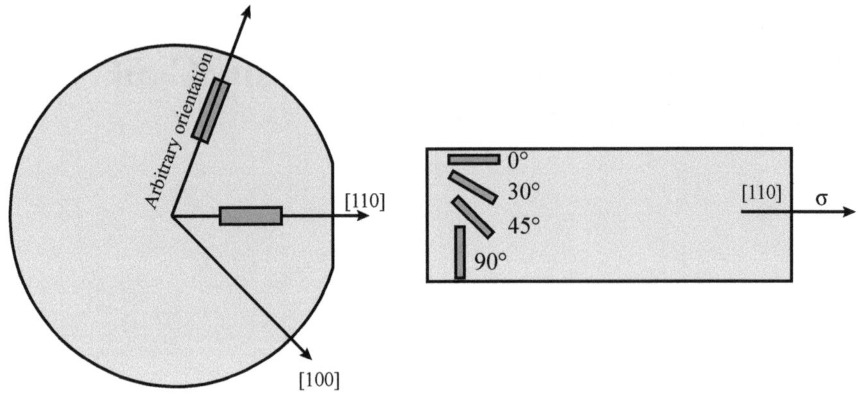

Fig. C.1 A schematic sketch of 3C-SiC piezoresistor fabricated on (100) plane, aligned in different orientations

The bending experiment was then carried out and the ratior between the relative resistance change with applied strain in [110] orientation ($[\Delta R/R]/\varepsilon$) was obtained. Figure C.3 plots the experimental data (the red dots) and the theoretical data calculated using Eq. C.3 (the blue curve). Evidently, experimental data was is a good agreement with theoretical estimation, indicating the feasibility of using the fundamental coefficients to predict the output of a SiC strain sensors aligned in an arbitrary orientation.

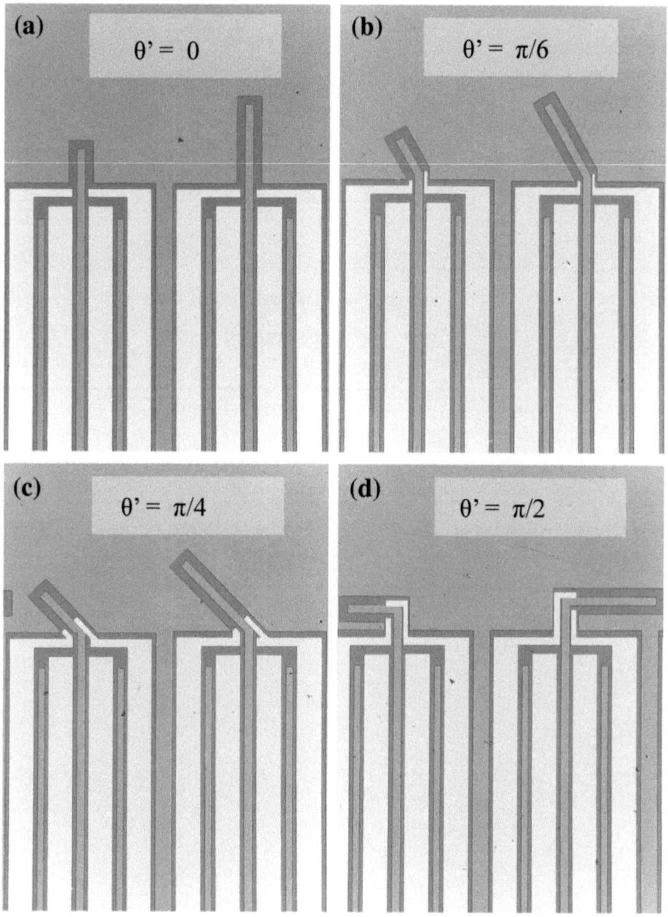

Fig. C.2 p-type 3C-SiC piezoresistors fabricated on (100) plane and aligned in different orientations

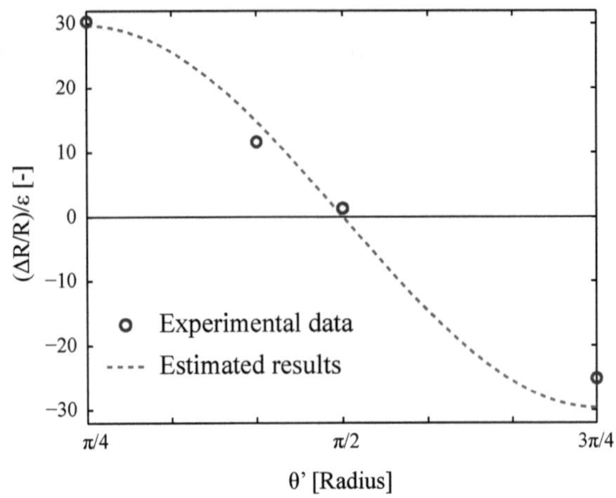

Fig. C.3 Comparison between the experimental data and the calculated results of the relative resistance change in p-type 3C-SiC aligned in different orientations under tensile strain pointed at [110] direction

Appendix D
Estimation of the GF of 3C-SiC Films with Different Thicknesses

Let ϕ_{hd} and ϕ_{ld} are the sheet conductance of the high density defect layer and the low density defect layer, respectively.

From Eq. 4.12, the conductance of the SiC film (G_t) is

$$G_t = G_{hd} + G_{ld} = \phi_{hd} t_{hd} + \phi_{ld} t_{ld} = \phi_{hd} t_{hd} + \phi_{ld}(t_{SiC} - t_{hd}) \tag{D.1}$$

Hence:

$$\frac{G_{hd}}{G_t} = \frac{\phi_{hd} t_{hd}}{\phi_{hd} t_{hd} + \phi_{ld}(t_{SiC} - t_{hd})} = f(t_{SiC}) \tag{D.2}$$

All SiC films are grown under the same conditions; therefore the high density defect layers are expected to have the same thickness (t_{hd} = constant) and the same properties. Hence, $f(t_{SiC}) = G_{hd}/G_t$ is a monotonically decreasing function of the film thickness t_{SiC} and $f(t_{SiC})$ has a value between 0 and 1.

The gauge factor is defined as:

$$GF = \frac{1}{\varepsilon} \times \frac{\Delta R}{R} \tag{D.3}$$

where R is the resistance of the SiC resistor; ε is the applied strain. The conductance of the SiC resistor is: $G = 1/R$. Therefore:

$$\begin{aligned} GF &= \frac{1}{\varepsilon} \times \frac{\Delta R}{R} \\ &= \frac{1}{\varepsilon} \times \frac{R_2 - R_1}{R_1} \\ &= \frac{1}{\varepsilon} \times \frac{\frac{1}{G_2} - \frac{1}{G_1}}{\frac{1}{G_1}} \\ &= \frac{1}{\varepsilon} \times \frac{G_1 - G_2}{G_2} \approx -\frac{1}{\varepsilon} \times \frac{\Delta G}{G} \end{aligned} \tag{D.4}$$

H.-P. Phan, *Piezoresistive Effect of p-Type Single Crystalline 3C-SiC*, Springer Theses, DOI 10.1007/978-3-319-55544-7

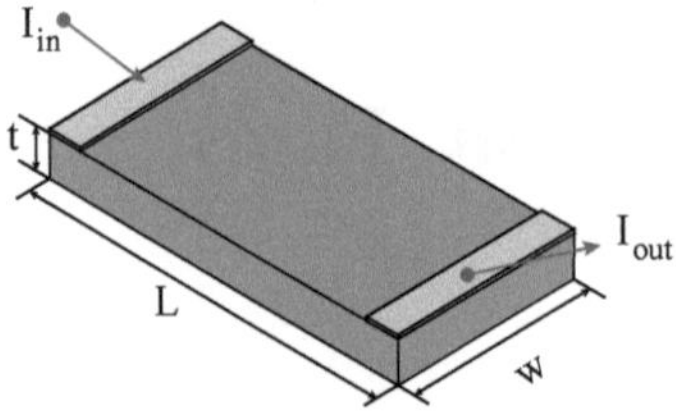

Fig. D.1 Concept of the conductance of a SiC film

Here, R_1 (or G_1) and R_2 (or G_2) are the resistance (or conductance) of the strain-free and strained SiC films, respectively. Note that $G_t = G_{ld} + G_{hd}$ and $\Delta G_t = \Delta G_{ld} + \Delta G_{hd}$, hence Eq. 4.15 in Chap. 4 can be deduced by:

$$\begin{aligned} GF &= -\frac{1}{\varepsilon} \times \frac{\Delta G_{ld} + \Delta G_{hd}}{G_t} \\ &= -\frac{1}{\varepsilon} \times \frac{\Delta G_{ld}}{G_{ld}} \times \frac{G_{ld}}{G_t} - \frac{1}{\varepsilon} \times \frac{\Delta G_{hd}}{G_{hd}} \times \frac{G_{hd}}{G_t} \\ &= GF_{ld}\frac{G_{ld}}{G_t} + GF_{hd}\frac{G_{hd}}{G_t} = (1 - f(t_{SiC}))GF_{ld} + f(t_{SiC})GF_{hd} \end{aligned} \tag{D.5}$$

where, GF_{ld} and GF_{hd} are the gauge factors of low density defect layer and high density defect layer, respectively.

$$\begin{cases} GF_{ld} = -\frac{1}{\varepsilon} \times \frac{\Delta G_{ld}}{G_{ld}} \\ GF_{hd} = -\frac{1}{\varepsilon} \times \frac{\Delta G_{hd}}{G_{hd}} \end{cases} \tag{D.6}$$

Estimation of the thickness dependence of the piezoresistive effect From the TEM image, the thickness of the defect layer is about 60 nm (t_{hd} = 60 nm). As 60 nm is much smaller than 1 μm, the influence of crystal defect in 1 μm can be negligible. Therefore the gauge factor of the 1 μm film is approximately equal to that of the low density defect layer ($GF_{ld} = 31.1$) (Fig. D.1).

For a quantitative demonstration, we have compared the the 280 and the 80 nm. For other thicknesses, the same method can be applied. Let μ and σ be the carrier mobility and conductivity of the SiC film. The mobility of the 80 and 280 nm film were measured to be approximately 7.5 and 15 cm^2/Vs, respectively. Therefore: $\mu_{280} = 2\mu_{80}$ The conductivity of the SiC film is:

$$\sigma = n\mu q \tag{D.7}$$

where n is carrier concentration, q is carrier charge. These films have the same doping concentration, therefore $\sigma_{280} = 2\sigma_{80}$.

The conductance of SiC film is:

$$G_t = \sigma w t / L \tag{D.8}$$

Hence

$$\sigma_{280} = \frac{G_{280}}{280}\frac{L}{w} = 2\sigma_{80} = 2\frac{G_{80}}{80}\frac{L}{w} \tag{D.9}$$

As the width and length of the films are the same, from Eq. D.9:

$$\frac{G_{280}}{280} = 2 \times \frac{G_{80}}{80} \tag{D.10}$$

Substituting Eq. D.1 into D.10:

$$\frac{\phi_{hd}60 + \phi_{ld}(280 - 60)}{280} = 2 \times \frac{\phi_{hd}60 + \phi_{ld}(80 - 60)}{80} \tag{D.11}$$

Hence, the ratio of the sheet conductance of the high density defect layer and the low density defect layer has been calculated as:

$$\frac{\phi_{hd}}{\phi_{ld}} = \frac{1}{4.5} \tag{D.12}$$

Substituting this value into Eq. D.2:

$$f(t_{SiC}) = \frac{t_{hd}}{t_{hd} + 4.5(t_{SiC} - t_{hd})} = \frac{60}{60 + 4.5(t_{SiC} - 60)} \tag{D.13}$$

Substituting Eq. D.12 into D.5, from the measured gauge factor of the 80 nm film, the gauge factor of the high density defect layer was calculated as $GF_{hd} = 4.1$ (Table D.1). From the values of G_{hd} and GF_{ld}, the estimated gauge factors of 3C-SiC films are listed in Table D.2.

Table D.1 Comparison between the estimated GF and measured GF

Thickness [nm]	Measured GF [-]	Estimated GF [-]
80	20.5	20.5
130	26.1	26.9
280	30.3	29.6
380	30.4	30.0
1000	31.1	30.5

Table D.2 List of symbols used in this appendix

Symbol	Definition
G_t	The total conductance of the SiC film
G_{hd}	The conductance of the high density defect layer of the SiC film
G_{ld}	The conductance of the low density defect layer of the SiC film
GF	The gauge factor of the SiC films
GF_{ld}	The gauge factor of the low density defect layer of the SiC film
GF_{hd}	The gauge factor of the high density defect layer of the SiC film
μ	The carrier mobility of the SiC film
σ	The conductivity of the SiC film
ϕ	The sheet conductance of the SiC film
ϕ_{hd}	The sheet conductance of high density defect layer of the SiC film
ϕ_{ld}	The sheet conductance of low density defect layer of the SiC film
t_{SiC}	The thickness of the SiC film
t_{hd}	The thickness of the high density defect layer of the SiC film
t_{ld}	The thickness of the low density defect layer of the SiC film
L	The length of the SiC resistor
w	The width of the SiC resistor

Appendix E
In Situ Characterization of the Strain Effect on p-Type 3C-SiC at High Temperatures

E.1 Measurement of Current Leak at SiC/Si When Using Joule Heating Effect

When increasing the temperature of SiC resistors using the Joule effect, the temperature of SiC resistor located at the center of the released bridge raised significantly, while the temperature at the vicinity of SiC/Si junction remained at approximately at room temperature. We measured the current leak at SiC/Si junction when increasing the electrical power applied to the SiC resistors. Evidently, at an applied voltage of 1.7 V, which corresponds to the temperature at SiC resistor of approximately 300 °C, the current leak through the SiC/Si heterojunction was below 20 nA, Fig. E.1. Note that, as the current through SiC resistor is about 200 μA, the leakage current is negligible.

E.2 Approximation of the Linear Relationship Between Applied Power and Resistance Change in Small Interval

From Fig. 4.22 in Chap. 4, we found that increasing the applied electrical power results in a decrease in the resistance of SiC, following the thermoresistive effect in p-type 3C-SiC. In the other words, the applied power is a monotonically decreasing function of SiC resistance: $P = f(R)$, which is a nonlinear function. However, in a sufficiently small interval of resistance change, we can approximate the applied power (P) as a linear function of R using Taylor's series approximation:

$$P = f(R) = f(R_0) + (R - R_0)\frac{f'(R_0)}{1!} + 0(R - R_0)^2 \tag{E.1}$$

H.-P. Phan, *Piezoresistive Effect of p-Type Single Crystalline 3C-SiC*,
Springer Theses, DOI 10.1007/978-3-319-55544-7

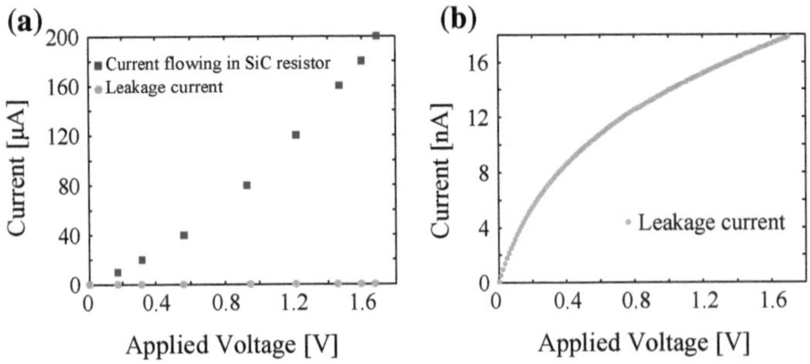

Fig. E.1 **a** Comparison of the current flowing in SiC resistors and the leakage current through the SiC/Si heterojunction when using the Joule heating effect. **b** The leakage current plotted in nA scale

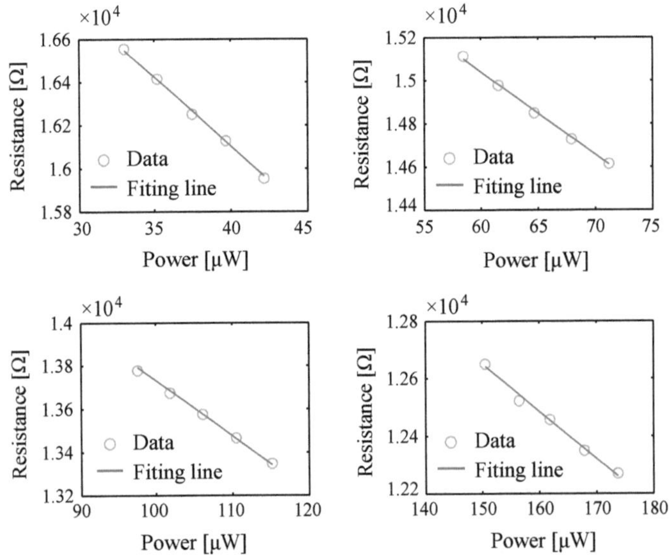

Fig. E.2 Experimental data showing the linear relationship between the applied power and resistance in a sufficiently small interval of resistance change (or of the applied power)

where $0(R - R_0)^2$ is the second order of Taylor's series, and in a small interval of R0 (or of the applied power), this factor is negligible. Our empirical results also indicated that in a sufficiently small interval of resistance change, the applied electrical power has a linear relationship with the resistance, as shown in Fig. E.2.

Appendix F
Orientation Dependence of the Piezoresistive Effect in p-Type 3C-SiC Four-Terminal Resistors

F.1 The Deduction of the Stress Tensors and Piezoresistive Coefficients

F.1.1 Stress Tensors

The ratio of the generated voltage (V_{out}) across terminals 3 and 4 to the applied voltage across terminals 1 and 2 (V_{in}) under stress can be quantified by the coefficients π_{61}, π_{62}, π_{63}, π_{64}, π_{65} and π_{66}, as expressed in the following equation [1]:

$$V_{out} = V_{in}(\pi_{61}\sigma_1 + \pi_{62}\sigma_2 + \pi_{63}\sigma_3 + \pi_{64}\sigma_4 + \pi_{65}\sigma_5 + \pi_{66}\sigma_6) \qquad (F.1)$$

where σ_i (i varying from 1 to 6) are the components of the stress tensors. For the case of a thin film semiconductor, the components related to the out-of-plane axis (π_{63}, π_{64}, and π_{65}) are negligible. Therefore, the ratio of the generated voltage (V_{out}) across terminals 3 and 4 to the applied voltage across terminals 1 and 2 (V_{in}) is [2]:

$$V_{out} = V_{in}(\pi_{61}\sigma_1 + \pi_{62}\sigma_2 + \pi_{66}\sigma_6) \qquad (F.2)$$

Let β be the angle between the direction of the current and applied stress, the normal stresses and shear stress can be obtained using Mohr's circle, as shown in Fig. F.1. Accordingly, the stress tensors are [3]:

$$\begin{cases} \sigma_1 = \sigma \cos^2 \beta \\ \sigma_2 = \sigma \sin^2 \beta \\ \sigma_6 = -\dfrac{\sin 2\beta}{2}\sigma \end{cases} \qquad (F.3)$$

H.-P. Phan, *Piezoresistive Effect of p-Type Single Crystalline 3C-SiC*,
Springer Theses, DOI 10.1007/978-3-319-55544-7

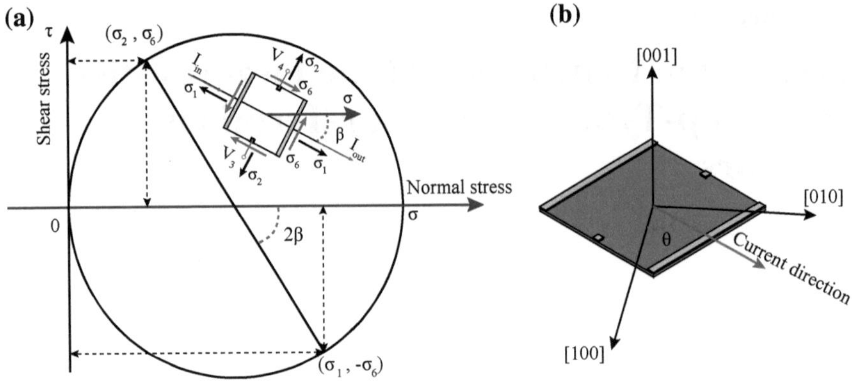

Fig. F.1 The Mohr's circle used to calculate the stress tensors applied to four terminal devices

F.1.2 Equation of Piezoresistive Coefficients

The piezoresistive coefficients π_{61}, π_{62}, and π_{66} are [2]:

$$\begin{cases} \pi_{61} = a(l_1^3 l_2 + m_1^3 m_2 + n_1^3 n_2) \\ \pi_{62} = -a(l_1^3 l_2 + m_1^3 m_2 + n_1^3 n_2) \\ \pi_{66} = b + 2a[(l_1 l_2)^2 + (m_1 m_2)^2 + (n_1 n_2)^2] \end{cases} \tag{F.4}$$

here, l_i, m_j, and n_k are direction cosines [4]. In (100) plane, when the coordinates rotate about [001] axis an angle of θ, the matrix $[l_i\ m_i\ n_k]$ is [4]:

$$[l\ m\ n] = \begin{bmatrix} \cos\theta & \sin\theta & 0 \\ -\sin\theta & \cos\theta & 0 \\ 0 & 0 & 1 \end{bmatrix}. \tag{F.5}$$

Substitute Eq. F.5 into F.4, we have:

$$\begin{cases} \pi_{61} = -a\dfrac{\sin 4\theta}{4} \\ \pi_{62} = a\dfrac{\sin 4\theta}{4} \\ \pi_{66} = b + a\sin^2 2\theta \end{cases} \tag{F.6}$$

From Eqs. F.2, F.3 and F.6, the ratio of the output voltage (V_{out}) and the input voltage (V_{in}) per unit applied stress is:

$$\frac{V_{out}}{V_{in}} = \pi_{61}\sigma_1 + \pi_{62}\sigma_2 + \pi_{66}\sigma_6 = \left[-\frac{a}{4}\sin 4\theta\cos^2\beta + \frac{a}{4}\sin 4\theta\sin^2\beta - \frac{1}{2}(b + a\sin^2 2\theta)\sin 2\beta\right]\sigma \tag{F.7}$$

References

1. D.V. Dao, T. Toriyama, J. Wells, S. Sugiyama, Silicon piezoresistive six-degree of freedom micro force-moment sensor. Sens. Mater. **15**, 113–135 (2002)
2. Y. Kanda, Graphical representation of the piezoresistance coefficients in silicon-shear coefficients in plane. Jpn. J. Appl. Phys. **26**(7), 1031 (1987)
3. R.C. Jaeger, J.C. Suhling, R. Ramani, Errors associated with the design, calibration and application of piezoresistive stress sensors in (100) silicon. IEEE Trans. Compon. Hybrids **17**(1), 97–107 (1994)
4. Y. Kanda, A graphical representation of the piezoresistance coefficients in silicon. IEEE Trans. Electron Devices **29**(1),64–70 (1982)

About the Author

Dr. Hoang-Phuong Phan received his B.Eng. and M.Eng. from The University of Tokyo, Japan in 2011 and 2013, and his Ph.D. from the Queensland Micro- and Nanotechnology Centre at Griffith University, Australia in 2016. Since 2016, he has been a research fellow at the Queensland Micro- and Nanotechnology Centre. He was also a visiting scholar at both the Advanced Institute of Science and Technology (AIST) and Aichi Institute of Technology (AIT) in Japan. His research interests include micro electromechanical systems (MEMS), advanced materials, physics of semiconductors, and nanotechnologies. Dr. Phan has published more than 30 peer-reviewed journal articles (e.g. Sci. Rep., IEEE Electron Device Lett., Appl. Phys. Lett., J. Mater. Chem., JMEMS, etc) and numerous conference papers (e.g. IEEE MEMS, Transducer, APCOT). He also holds two US patents (pending) on MEMS pressure sensors and acoustic transmitters. Dr. Phan was a recipient of Japanese Government Scholarships (MEXT) for both undergraduate and postgraduate studies, the Griffith University postgraduate scholarship, and the international student scholarship for Ph.D. course. Dr. Phan was awarded the Springer Outstanding Thesis award and selected to attend the 23rd World Micromachines Summit in 2017 attended by the Australian delegates.

Publications

1. **H.-P. Phan**, T. Dinh, T. Kozeki, T.-K. Nguyen, A. Qamar, T. Namazu, N.-T. Nguyen, and D. V. Dao, "Nano strain-amplifier: making ultra-sensitive piezoresistance in nanowires possible without the need of quantum and surface charge effects," *Appl. Phys. Lett.*, vol. 109, p. 123502, 2016. IF: 3.142
2. **H.-P. Phan**, T. Dinh, T. Kozeki, A. Qamar, T. Namazu, S. Dimitrijev, N.-T. Nguyen, and D. V. Dao, "Piezoresistive effect in p-type 3C-SiC at high temperatures characterized using Joule heating," *Scie. Rep.*, vol. 6, no. 1–10, p. 28499, 2016

H.-P. Phan, *Piezoresistive Effect of p-Type Single Crystalline 3C-SiC*,
Springer Theses, DOI 10.1007/978-3-319-55544-7

3. H.-P. Phan, D. V. Dao, K. Nakamura, S. Dimitrijev, and N.-T. Nguyen, "The Piezoresistive Effect of SiC for MEMS Sensors at High Temperatures: A Review," *IEEE/ASME Journal of Microelectromechanical Systems (JMEMS)*, vol. 24, no. 6, p. 1663–1677, 2015
4. H.-P. Phan, T. Kozeki, T. Dinh, A. Qamar, Y. Zhu, T. Namazu, N.-T. Nguyen, and D. V. Dao, "Piezoresistive effect of p-type silicon nanowires fabricated by a top-down process using FIB implantation and wet etching," *RSC Adv.*, vol. 5, pp. 82121–82126, 2015
5. H.-P. Phan, D. V. Dao, P. Tanner, L. Wang, N.-T. Nguyen, Y. Zhu, and S. Dimitrijev, "Fundamental piezoresistive coefficients of p-type single crystalline 3C-SiC," *Appl. Phys. Lett.*, vol. 104, no. 11, p. 111905, 2014
6. I. Shimoyama, K. Matsumoto, B.-K. Nguyen, M.-D. Nguyen, and H.-P. Phan, Pressure-Sensitive Sensor, US Patent (Appl.), 20150362394, 2015

MIX
Papier aus verantwortungsvollen Quellen
Paper from responsible sources
FSC® C105338

If you have any concerns about our products,
you can contact us on
ProductSafety@springernature.com

In case Publisher is established outside the EU,
the EU authorized representative is:
Springer Nature Customer Service Center GmbH
Europaplatz 3, 69115 Heidelberg, Germany

Printed by Libri Plureos GmbH
in Hamburg, Germany